•◆•

The publication of this book is sponsored by
Cornell University, the original home of
The Physical Review.

It is jointly published by
The American Physical Society and
the American Institute of Physics
in celebration of the 100th anniversary of
The Physical Review.

•◆•

Library of Congress Cataloging-in-Publication Data
Hartman, Paul, 1913 –
  A memoir on The physical review : a history of the first hundred
years / Paul Hartman.
      p. cm.
  ISBN 1-56396-282-9 (pbk. : alk. paper)
  1. Physical review—History. 2. Physical review letters—History.
I. Title. II. Title: Physical review.
QC1.H29 1994   94-5660
530' .05—dc20

10 9 · 8 7 6 5 4 3 2 1

©1994 by Cornell University.
Published by American Institute of Physics.
All rights reserved.
Printed in the United States of America.

AIP Press
American Institute of Physics
500 Sunnyside Boulevard
Woodbury, NY 11797-2999

Library of Congress Cataloging-in-Publication Data
Hartman, Paul, 1913 –
   A memoir on The physical review : a history of the first hundred
years / Paul Hartman.
      p. cm.
   ISBN 1-56396-282-9 (pbk. : alk. paper)
   1. Physical review—History.   2. Physical review letters—History.
I. Title. II. Title: Physical review.
QC1.H29 1994   94-5660
530' .05—dc20

10 9 8 7 6 5 4 3 2 1

Contents

•◆•

Contents

Foreword

•◆•

Two years ago we at APS were beginning to prepare ourselves for the celebration of the centennial year of the founding of the PHYSICAL REVIEW at Cornell University. We considered a number of possible activities, among them the commissioning of a volunteer to prepare a history of our distinguished journals to appear sometime during the centennial year. Imagine our pleasure in hearing that indeed such a "history" had already been written, by a person who was in as good a position as anyone to accomplish this difficult task. Paul Hartman had been a professor in the Cornell Physics Department for many years, and while now officially retired still kept an office there. He had started this project perhaps eight years ago. By the time I saw the finished work it had been refined and updated, and the final result is the volume you are now holding. And it should be noted that Professor Hartman had a more modest goal in mind than in the preparation of a scholarly work—he appropriately calls it a "memoir" rather than a history, to indicate (as he carefully warns us) that this is more a labor of love than an attempt to prepare a definitive chronicle of the journal's past glory. The true "history" of PHYSICAL REVIEW lies in the articles it has published; thus one might consider this book a companion to one containing an ambitious collection of seminal articles that have appeared in PHYSICAL REVIEW (and PHYSICAL REVIEW *Letters*) since its inception. The latter book is being jointly published at this time by the American Physical Society and the American Institute of Physics, under the editorship of Professor Henry Stroke, with the active assistance of fifteen distinguished American physicists.

The present volume is a gold mine for afficionados of what until very recently was known as the "jolly green giant," or by perhaps other not so flattering appellations, although of course in the years which are the ones mainly covered in this Memoir the physical size of the journal was far less daunting and more readily accessible to most physicists, specialties notwithstanding. Afficionados aside, there is much in this work of interest to all of us who publish, or have been published in

PHYSICAL REVIEW. Speaking for The American Physical Society I am happy to express my appreciation to Professor Hartman for his dedication and skill in creating this work.

I would also like to thank the Cornell University Physics Department for a generous contribution which enabled us to produce this volume, and to both APS and AIP for supplementing the Cornell contribution. Finally I thank Maria Taylor, Publisher of AIP Press, for all her help in getting this volume into print.

Benjamin Bederson
New York University

Preface

❖❖

One hundred years ago, mid-summer 1893, there appeared a modest new scientific publication devoted to physics, sponsored by the Physics Department of a young and still small university in upstate New York. The new journal was called the PHYSICAL REVIEW, and the university was Cornell, then itself founded but twenty-five years earlier. Today the sponsors of the journal and the founders of the university would not recognize their respective creations. The PHYSICAL REVIEW has become the most prestigious, manifold, and voluminous journal in American physics, if not world physics, and the university, now large, one of the leading world institutions of learning. The once small Physics Department is itself large and widely recognized.

In April of 1993, The American Physical Society sponsored a broadly based symposium to commemorate the centenary of the PHYSICAL REVIEW. In October of the year, *Physics Today*, in a special issue, celebrated the REVIEW. As further contribution, it had been suggested at Cornell some years before that some sort of retrospection of the PHYSICAL REVIEW be written. This book attempts to make that contribution.

I make no claim that I should be the one to undertake such an account; there are many in the physics community who could do it much better—people better at writing, with more erudition, comfortable with history, and more knowledgeable of the REVIEW and of physics. The slight justification for my undertaking it was the thought that having written a sort of "history" of the Cornell Physics Department, with which the PHYSICAL REVIEW was inextricably linked during its first two decades, it might come somewhat naturally to continue with a sort of "history" of the REVIEW. It is not clear that this necessarily follows. The previous writing was aimed parochially at those who had connection with the Cornell Physics Department; as such it could be rather familiar and anecdotal in tone. A history of the PHYSICAL REVIEW should obviously be intended to reach a larger audience, one beyond that having a Cornell connection. Whether it can then be made interesting and agreeable enough to hold a reader's attention is a judgment I can't make. It is also true that I have not been closely associated with the journal, while fifty years association with its founding Department lent some credibility to what I wrote concerning the latter. What will be found herein concerning the first Editors will have been largely "lifted" from that earlier endeavor.

It may be objected that I have included too much which is mainly of interest to Cornellians, in the bit of Cornell pre-REVIEW history presented, and in the short vignettes of the first Editors sketched. Such may not be immediate to a history of the PHYSICAL REVIEW but, it seems to me, does serve to round out the times and to put some flesh on the characters involved. So also with some of the correspondence quoted, and in sundry trivia and comments offered at various points, an attempt to bring the past state of physics to the reader's mind and possibly to bring a little life to what may otherwise be somewhat dull going. If I have been too overbearing in these matters and unsuccessful in the whole, I plead guilty and I'm sorry. As is often done in physics texts with sections of material not crucial to subsequent subject matter, the reader is invited to skip those sections (here on Cornell and on the first Editors) and go on to the main course.

In its December 1989 issue, the *American Journal of Physics* published a nice article by Melba Phillips surveying the first fifty years of The American Physical Society. Pretty clearly some of my "asides" are prime events in her survey, e.g., the founding. Since the progress of the PHYSICAL REVIEW and the doings of the Society have been so intimately tied together, I have chosen herein not to delete what I had put down of the Society affairs. What she has written is in general more comprehensive than what will be found in my story, and is well worth the reading. I hope that some duplication between the two histories will not be detrimental to either one.

Somewhat the same may be said of the special issue of *Physics Today* celebrating the REVIEW centennial. I have not altered my manuscript for it, so there is clearly going to be some duplication here as well. So be it.

Already, in these first few paragraphs we have come on a problem which will be with us throughout what is to follow. It is one of form and accuracy. The official title of our journal is THE PHYSICAL REVIEW. In the text to come, while "PHYSICAL REVIEW" will be arbitrarily capitalized, the word "THE" will not always be done so and, in the interest of readability, frequently will not even be used. Indeed, as often as not, the word "PHYSICAL" will be omitted and the capitalized "REVIEW" simply be used, which should occasion no confusion to the reader as to the journal's true name. It is a mode taken now and then by our early Editors, seen in at least a couple of notices quoted later. In many instances, mostly in parenthetical reference, the abbreviation *Phys. Rev.* will be employed. When we come to PHYSICAL REVIEW *Letters* (and *Abstracts*), there is no article "The" in the official title(s); we will italicize "*Letters*" ("*Abstracts*") when referring to the journal(s). The abbreviation *Phys. Rev. Lett.* will be used, and here and there simply *Letters*.

In no way does the record to follow masquerade as anything like good history. My "research" and peripheral reading related to the journal has been rather casual, my sources primarily local, and what has been learned is not properly referenced, if at all. There is absolutely no comparison to be made, for instance, to the exhaustive, yet engrossing, deeply researched, meticulously referenced, and well written work of Daniel Kevles in his *The Physicists*. (We will be borrowing here and there from that first rate volume.) Nor will we pretend to make anything like as significant a contribution as Kevles has made.

Nor will we be attempting to emulate the also scholarly and well referenced article of historian John W. Servos: "A Disciplinary Program that Failed: Wilder D. Bancroft and the *Journal of Physical Chemistry*, 1896–1933" [*Isis*, **73**, 207–232 (1982)]. As interesting but far more limited than Kevles, it is an insightful tale with characters and setting rather closer to "home" than is found in *The Physicists*. If there has been serious dissension and in-fighting connected with THE PHYSICAL REVIEW as there was with the *Journal of Physical Chemistry*, it has not come to this "writer's" attention, and he is not prepared, nor has he the temperament or ability to do the research to dig any such out.

Any comparison, therefore, between what follows and treatments like the foregoing pair would be ridiculous. Nevertheless, the occasion we are celebrating calls for at least some sort of lengthy review of *the* REVIEW, inadequate though it may be. But perhaps something suitable can be done for it here. We'll see.

Paul Hartman
Cornell University

Acknowledgments

•◆•

This project originated in the mind of Neil Ashcroft who, looking ahead to 1993, felt some kind of celebration and retrospection on the REVIEW was necessary. Whether this be that retrospection or not, I appreciate his encouragement in the enterprise and thank him for needling me into it. It turned out to be rather interesting. Thanks also should go to Albert Silverman, Dale Corson, Carl Franck and others, besides Neil, who have looked at it and given their opinions about its usefulness. In a critical reading of the manuscript, Professor David Lazarus, former Editor-in-Chief of the PHYSICAL REVIEW, was especially helpful in making many corrections, and offering constructive comments and suggestions, not all of which have been taken. Present Editor-in Chief, Ben Bederson, was similarly helpful. Spencer Weart and Melba Phillips at AIP have seen it and given their opinions. At Cornell, the Physical Sciences Library, the Archives of the University Library, and Physics Department holdings made the whole thing possible. Ben Widom and Richard Zare were helpful in informing me on Bancroft and the *Journal of Physical Chemistry* story of John Servos. Jim Krumhansl and Bob Richardson were helpful in our deciding what to do with the manuscript. Liane Cooper and others on the secretarial staff of the Laboratory of Atomic and Solid State Physics did the final typing of the manuscript. Connie Wright did the laborious assembly and proof corrections of the whole. Thanks is owed to all who have been involved one way or another.

Special appreciation is extended to Cornell, to The American Physical Society, and the American Institute of Physics, which together have made publication possible.

Paul Hartman

Illustrations

···

FOUNDING
FATHERS

•◆•

Edward L. Nichols, founder of the PHYSICAL REVIEW (1893) and a first editor (1893–1913) until the journal was taken over by The American Physical Society. He was head of the Cornell Physics Department from 1887 to 1919, and from 1907 to 1909 served as the president of the APS. This portrait appeared with the Memorial Statement of Nichols written by Professor Merritt in the first issue of the 1938 journal.

Edward L. Nicholas
(1854 •◆• 1937)

Ernest Merritt, second of the trio of first editors; his editorship lasted two decades. He succeeded Nichols as Cornell Physics Department head in 1919 and served until his own retirement in 1935. He served the APS in various capacities for nearly fifty years, including the positions of first secretary and president (1914–1916).

Ernest Merritt
(1865 ·◆· 1948)

Frederick Bedell, third of the editorship trio. After the APS acquired the journal in 1913, he continued as managing editor for another decade (1913–1923) supported by help of the editorial board. This photograph (courtesy of Dr. Caroline Bedell Thomas) is taken during that period.

Frederick Bedell
(1868 ·◆· 1958)

John Torrence Tate
(1889 •◆• 1950)

John Torrence Tate, the last long-time (1926–1950) Review *editor (Managing Editor) in the Pre-World War II period. He was president of the APS in 1939 and one of the founders of the American Institute of Physics. The portrait accompanied the Memorial Statement for him in the first issue of the 1950 journal.*

Edward L. Nicolas, Lord Kelvin,
and J. Gould Sherman

*Edward L. Nichols, Lord Kelvin, and J. Gould Schurman (Cornell President)
outside Cornell's dynamo laboratory on the occasion of Lord Kelvin's visit to
Cornell. Schurman may have provided the incentive for Nichols to establish an
American journal of physics.*

Second Series   *May, 1923*   *Vol. 21, No. 5*

THE
PHYSICAL REVIEW

A QUANTUM THEORY OF THE SCATTERING OF X-RAYS BY LIGHT ELEMENTS

By Arthur H. Compton

Abstract

A quantum theory of the scattering of X-rays and γ-rays by light elements. —The hypothesis is suggested that when an X-ray quantum is scattered it spends all of its energy and momentum upon some particular electron. This electron in turn scatters the ray in some definite direction. The change in momentum of the X-ray quantum due to the change in its direction of propagation results in a recoil of the scattering electron. The energy in the scattered quantum is thus less than the energy in the primary quantum by the kinetic energy of recoil of the scattering electron. The corresponding *increase in the wave-length of the scattered beam* is $\lambda_\theta - \lambda_0 = (2h/mc) \sin^2 \frac{1}{2}\theta = 0.0484 \sin^2 \frac{1}{2}\theta$, where h is the Planck constant, m is the mass of the scattering electron, c is the velocity of light, and θ is the angle between the incident and the scattered ray. Hence the increase is independent of the wave-length. *The distribution of the scattered radiation* is found, by an indirect and not quite rigid method, to be concentrated in the forward direction according to a definite law (Eq. 27). The total energy removed from the primary beam comes out less than that given by the classical Thomson theory in the ratio $1/(1 + 2\alpha)$, where $\alpha = h/mc\lambda_0 = 0.0242/\lambda_0$. Of this energy a fraction $(1 + \alpha)/(1 + 2\alpha)$ reappears as scattered radiation, while the remainder is truly absorbed and transformed into kinetic energy of recoil of the scattering electrons. Hence, if σ_0 is the *scattering absorption coefficient* according to the classical theory, the coefficient according to this theory is $\sigma = \sigma_0/(1 + 2\alpha) = \sigma_s + \sigma_a$, where σ_s is the true scattering coefficient $[(1 + \alpha)\sigma/(1 + 2\alpha)^2]$, and σ_a is the coefficient of absorption due to scattering $[\alpha\sigma/(1 + 2\alpha)^2]$. Unpublished experimental results are given which show that for graphite and the Mo–K radiation the scattered radiation is longer than the primary, the observed difference $(\lambda_{\pi/2} - \lambda_0 = .022)$ being close to the computed value .024. In the case of scattered γ-rays, the wave-length has been found to vary with θ in agreement with the theory, increasing from .022 A (primary) to .068 A ($\theta = 135°$). Also the velocity of secondary β-rays excited in light elements by γ-rays agrees with the suggestion that they are recoil electrons. As for the predicted variation of absorption with λ, Hewlett's results for carbon for wave-lengths below 0.5 A are in excellent agreement with this theory; also the predicted concentration in the forward direction is shown to be in agreement with the experimental results,

Title page of the Physical Review, *Volume 21, 1923. This paper by Arthur H. Compton covering the discovery of the important quantum effect bearing his name was probably the most important paper that the* Review *had published up to this time.*

scattering electron in motion at an angle of less than 90° with the primary beam. But it is well known that the energy radiated by a moving body is greater in the direction of its motion. We should therefore expect, as is experimentally observed, that the intensity of the scattered radiation should be greater in the general direction of the primary X-rays than in the reverse direction.

The change in wave-length due to scattering.—Imagine, as in Fig. 1A,

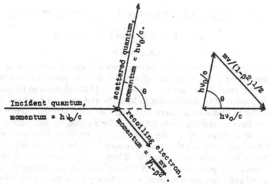

Fig. 1 A   Fig. 1 B

that an X-ray quantum of frequency ν_0 is scattered by an electron of mass m. The momentum of the incident ray will be $h\nu_0/c$, where c is the velocity of light and h is Planck's constant, and that of the scattered ray is $h\nu_\theta/c$ at an angle θ with the initial momentum. The principle of the conservation of momentum accordingly demands that the momentum of recoil of the scattering electron shall equal the vector difference between the momenta of these two rays, as in Fig. 1B. The momentum of the electron, $m\beta c/\sqrt{1-\beta^2}$, is thus given by the relation

$$\left(\frac{m\beta c}{\sqrt{1-\beta^2}}\right)^2 = \left(\frac{h\nu_0}{c}\right)^2 + \left(\frac{h\nu_\theta}{c}\right)^2 + 2\frac{h\nu_0}{c}\cdot\frac{h\nu_\theta}{c}\cos\theta, \qquad (1)$$

where β is the ratio of the velocity of recoil of the electron to the velocity of light. But the energy $h\nu_\theta$ in the scattered quantum is equal to that of the incident quantum $h\nu_0$ less the kinetic energy of recoil of the scattering electron, i.e.,

$$h\nu_\theta = h\nu_0 - mc^2\left(\frac{1}{\sqrt{1-\beta^2}} - 1\right). \qquad (2)$$

We thus have two independent equations containing the two unknown quantities β and ν_θ. On solving the equations we find

$$\nu_\theta = \nu_0/(1 + 2\alpha\sin^2\tfrac{1}{2}\theta), \qquad (3)$$

A page from the beginning of Compton's paper showing the diagram and equations now familiar to every physicist.

Second Series   *December, 1927*   *Vol. 30, No. 6*

THE
PHYSICAL REVIEW

DIFFRACTION OF ELECTRONS BY A CRYSTAL OF NICKEL

By C. Davisson and L. H. Germer

Abstract

The intensity of scattering of a homogeneous beam of electrons of adjustable speed incident upon a single crystal of nickel has been measured as a function of direction. The crystal is cut parallel to a set of its {111}-planes and bombardment is at normal incidence. The distribution in latitude and azimuth has been determined for such scattered electrons as have lost little or none of their incident energy.

Electron beams resulting from diffraction by a nickel crystal.—Electrons of the above class are scattered in all directions at all speeds of bombardment, but at and near critical speeds sets of three or of six sharply defined beams of electrons issue from the crystal in its principal azimuths. Thirty such sets of beams have been observed for bombarding potentials below 370 volts. Six of these sets are due to scattering by adsorbed gas; they are not found when the crystal is thoroughly degassed. Of the twenty-four sets due to scattering by the gas-free crystal, twenty are associated with twenty sets of Laue beams that would issue from the crystal within the range of observation if the incident beam were a beam of heterogeneous x-rays, three that occur near grazing are accounted for as diffraction beams due to scattering from a single {111}-layer of nickel atoms, and one set of low intensity has not been accounted for. *Missing beams* number eight. These are beams whose occurrence is required by the correlations mentioned above, but which have not been found. The intensities expected for these beams are all low.

The spacing factor concerned in electron diffraction by a nickel crystal.—The electron beams associated with Laue beams do not coincide with these beams in position, but occur as if the crystal were contracted normally to its surface. The spacing factor describing this contraction varies from 0.7 for electrons of lowest speed to 0.9 for electrons whose speed corresponds to a potential difference of 370 volts.

Equivalent wave-lengths of the electron beams may be calculated from the diffraction data in the usual way. These turn out to be in acceptable agreement with the values of h/mv of the undulatory mechanics.

Diffraction beams due to adsorbed gas are observed except when the crystal has been thoroughly cleaned by heating. Six sets of beams of this class have been found; three of these appear only when the crystal is heavily coated with gas; the other three only when the amount of adsorbed gas is slight. The structure of the gas film giving rise to the latter beams has been deduced.

T HE investigation reported in this paper was begun as the result of an accident which occurred in this laboratory in April 1925. At that time we were continuing an investigation, first reported in 1921,[1] of the distribution-in-angle of electrons scattered by a target of ordinary (poly-

---

[1] Davisson & Kunsman, Science **64**, 522, (1921).

705

Title page of the Physical Review, *Volume 30, 1927. While the discovery of electron waves had been announced in* Nature, *the* Review *contained the first detailed description of the famous experiment and its results.*

Droplet Fission of Uranium and Thorium Nuclei

The Fifth Washington Conference on Theoretical Physics, sponsored jointly by George Washington University and the Carnegie Institution of Washington, began January 26, 1939, with a discussion by Professor Bohr and Professor Fermi of the remarkable chemical identification by Hahn and Strassmann in Berlin of radioactive barium in uranium which had been bombarded by neutrons. Professors Bohr and Rosenfeld had brought from Copenhagen the interpretation by Frisch and Meitner that the nuclear "surface-tension" fails to hold together the "droplet" of mass 239, with a resulting division of the nucleus into two roughly equal parts. Frisch and Meitner had also suggested the experimental test of this hypothesis by a search for the expected recoil-particles of energies well above 100,000,000 electron-volts which should result from such a process. The whole matter was quite unexpected news to all present.

We immediately undertook to look for these extremely energetic particles, and at the conclusion of the Conference on January 28 were privileged to demonstrate them to Professors Bohr and Fermi. It was subsequently learned that the particles had been observed independently by Fowler and Dodson at Johns Hopkins the same day, by Dunning and co-workers at Columbia on January 25, and by Frisch in Copenhagen two weeks earlier.

For observations of the high energy particles, an ionization-chamber, about five mm deep, was placed about three cm below the neutron-source and was so arranged that interchangeable copper disks about three cm in diameter could be placed on the collector, which was connected to a linear pulse-amplifier. The upper faces of these disks were then coated with the materials to be tested.

With the amplifier feeding a cathode-ray oscillograph the usual alpha-particle pulses were observed when a layer of uranium oxide was placed on the disk. On exposure to

neutron-radiation from (Li+D) at 1000 kv two additional groups of pulses were observed. The first group corresponded to the "neutron-recoils" from the air in the chamber, as previously measured with the same amplifier gain and without the uranium. These neutron-recoils gave pulses about four times the size of the alpha-particle pulses. The second additional group was 20 to 40 times larger than the largest "recoil"-pulse, thus corresponding to energies of 75 to 150 Mev released in the chamber, or 150 to 300 Mev total energy for each individual process. With paraffin surrounding source and chamber the yield was roughly 30 counts per min. per μA of 1000-kv deuterons, which is a neutron-intensity corresponding to about 10,000 millicuries of radon-beryllium. The yield from thorium was of the same order of magnitude.

No effect was observed from bismuth, lead, thallium, mercury, gold, platinum, tungsten, tin or silver with as much as 1/1000 the intensity of that from uranium and thorium.

No effect was observed with either uranium or thorium produced by the gamma-rays from 3 μA of 1000-kv protons on lithium or on fluorine.

To determine roughly the energy-range of the neutrons involved in the fission-process, observations were made with the neutrons from several reactions, both with and without cadmium surrounding the ionization-chamber to filter out the thermal neutrons produced in the surrounding paraffin. Bearing in mind that the ratio of the counts with cadmium and without cadmium depends to a large extent on the amount of paraffin surrounding the source and chamber, the results of these tests may be deduced from Table I in which the relative number of "fissions" is given, with the total yield for uranium and thorium with high energy neutrons, being approximately equal, taken as 100 on an arbitrary scale.

From these comparisons it appears that the uranium fissions are produced by different processes for fast and slow neutrons, the fast-neutron process requiring more than 0.5 Mev but less than 2.5 Mev for effective operation. For thorium, on the other hand, only the fast-neutron process is effective, but somewhat surprisingly it also appears to require between 0.5 and 2.5 Mev.

R. B. Roberts
R. C. Meyer
L. R. Hafstad

Department of Terrestrial Magnetism,
Carnegie Institution of Washington,
Washington, D. C.,
February 4, 1939.

TABLE I.

NEUTRON-REACTION	MAXIMUM NEUTRON-ENERGY	URANIUM		THORIUM	
		No Cd	WITH Cd	No Cd	WITH Cd
	Mev				
Li+D	13.5	100	70	100	100
D+D	2.5	100	70	100	100
C+D	0.5	100	10	0	0

A letter to the editor in the PHYSICAL REVIEW, Volume 55, 1939 (February 15 issue). The first communication on fission to appear in the journal was followed by several others in the same issue, including one by Niels Bohr. Bohr's letter was submitted, as much as anything, to rectify his inadvertent disclosure of information and to establish that his European colleagues, Otto Frisch and Lise Meitner, had prior explanations for the phenomenon.

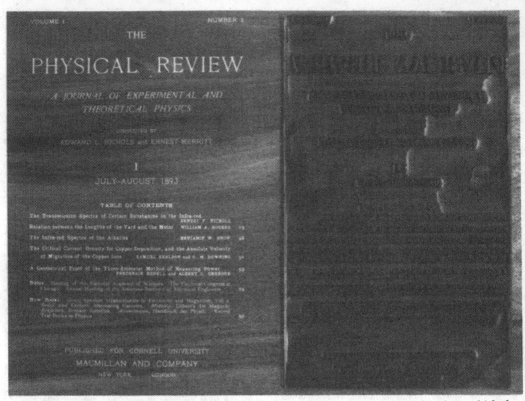

Plaque with a reproduction of the first cover of the PHYSICAL REVIEW and the lead plate from which the cover was printed. This plaque hangs on the wall in the office of physicist and Cornell President Emeritus, Dale Corson.

Ithaca, N. Y.,   May 3/93.

Professor E.L.Nichols,
        Ithaca, N.Y.

Dear Sir:-
        The Executive Committee yesterday appropriated $500.

for the Physics Review.

                    Yours truly, C. L. Williams
                                        Sec'y.

Memo from the Cornell Treasurer's Office notifying Nichols of the $500 appropriation to the journal made by the Board of Trustees. Memo is signed by E. L. Williams.

A history of the first hundred years

◆

INTRODUCTION

·◆·

Attics of physics laboratories are frequently repositories of old discarded apparatus, physics memorabilia, reprints of department publications, assorted junk, and worn out furniture given up by various professors, alive and deceased. Cornell's Rockefeller Hall, long time home of its Physics Department, has been no exception to this observation. By the end of World War II the Rockefeller attic was a veritable treasure house, quite in disarray, a conglomeration of stuff which has been likened to that in Tutankhamen's tomb at its opening; material had been accumulating there from the time of the building's occupancy forty years earlier. Anyone starting a new research project in the Department would head for the attic to find what he or she needed to get going. In particular, in 1946, or thereabouts, while prospecting for lead with which to make shielding for high energy radiations emanating from the 300 MeV synchrotron then being constructed (in parallel with Macmillan's machine at Berkeley) by the high energy physicists in the Department, the prospectors headed for the attic. It was well known that the Department, financed by the University, for twenty years had published the PHYSICAL REVIEW, today the "flagship" of the extensive list of publications served or published by the American Institute of Physics; one knew that up in the attic were large collections of the journal and artifacts associated with its publication. There were lead plates stacked around from which were printed figures, text, and contents of early copies of the journal. The high energy physicists' search was successful in a way not anticipated. True, much lead was uncovered (some 3000 pounds) and subsequently melted down into lead bricks, some of which are undoubtedly still playing their role in Cornell high energy physics. In addition, however, two plates were discovered which are now museum pieces hanging on two physicists' office walls. One is the lead plate from which was printed the cover page of Number 1, Volume I, July–August 1893, of the PHYSICAL REVIEW, the plate now nicely mounted in President Emeritus Dale Corson's office in Clark Hall. The other plate saved is that from which was

1

printed the first page of the same issue, that plate now gracing
Robert Wilson's office somewhere. One hopes that one day these
will wind up side by side in an established physics museum.

The flyleaf of the first volume of the journal as bound in the
Science Library at Cornell says it all:

THE
PHYSICAL REVIEW
A Journal of Experimental and
Theoretical Physics
Conducted by
Edward L. Nichols
and
Ernest Merritt
Published for Cornell University
THE MACMILLAN COMPANY
New York   London
Berlin: Mayer & Muller
1894

On the back of the page is a notice by Macmillan that the material is
copyrighted, and at the bottom that the printing is by:

Norwood Press
J. S. Cushing & Co.   Berwick & Smith
Boston, Massachusetts, USA

Starting with Volume V, however, the printing would be done by the
New EraPrinting Company of Lancaster, Pennsylvania. There must
have been a malfunction in the Norwood operation, for a letter found
from the editor of *Science,* James Catell, to the REVIEW, recommends
that the printing be done by the New Era Press. Since then typogra-
phers of Lancaster should have become rather knowing of physics.
Over much of its life, the PHYSICAL REVIEW was printed in that
locality.

While it is no longer "conducted by" Edward L. Nichols and
Ernest Merritt (and Frederick Bedell, who would join them shortly)
or published by Macmillan, it still is, as it maintains, "A Journal of
Experimental and Theoretical Physics."

(The cover page plate in Corson's office carries the same information as above but in different format, with a listing of the contents
below. The bound Volume One of the journal in the Cornell Science

Library carries a flyleaf as described above, followed by a page or so with alphabetical listing of papers and journal departments for the whole year, with no contents pages for each issue. A rather complete index for the year is at the back of the volume at the end of Number 6. Presumably each issue had a cover like the first one; covers are not bound in the Cornell collection of these early volumes, which are rather handsomely done, even to gilding page edges.)

Roughly forty years after this first expedition to the Rockefeller Hall attic, there was another one, similarly successful—this time by a contingent of chemists. (What were *they* doing in a Physics attic?) Back in a far, dark corner, they opened up a large wooden chest. Inside was a treasure—a lot of records having to do with the early PHYSICAL REVIEW: volumes of correspondence, lists of subscribers, billing and receipt forms, and the like; unfortunately nothing to do with the incentives for starting the journal. Nonetheless, it was a nice discovery made by our chemist friends. There it had been all those years; over future years it will reside in the Cornell Library Archives.

In another quarter of the attic were stored many sets of Series I of the REVIEW; the Physics Department for years had a small yearly income from the sale of nearly complete sets up to a certain year. "Nearly complete sets" is said advisedly. The noted radiologist, Lauritson Taylor, tells how when he was a student in the Department, Professor Merritt, then Department Chairman, offered him the job, at 25¢ an hour, of straightening out the chaos in the back issues held by the Department. They were in great disordered piles in the "dynamo" laboratory (as will be seen, Cornell physics was big in dynamo technology). After the job was finished, Taylor was to be given a complete set of the back issues; Merritt promised him that. So he went at it and after earning a considerable number of 25¢ pieces, finished the job. But in it all there was only a single complete set. On reporting this to Merritt, Taylor was told: "Take it. It's yours." The rest went up to the attic repository to be sold over the years to this institution or that.

This is not to imply that there are not other complete sets of Series I. Certainly libraries (including Cornell's) have them (not necessarily in the original) and the first editors had them. Where the latter sets have gone is not known. Taylor is not sure where his set wound up; not in his study at any rate. George Platczek occupied Professor Bedell's old office in Rockefeller Hall in the late 1930's before the old man's quarters were cleared of his books. There were all those PHYSICAL REVIEW's lining the shelves staring down on him. (George could be glad it wasn't 1980!) Rose Bethe has told how in low moments, George said he "felt the ghost of Frederick Bedell haunting him there."

Taylor's task was fairly formidable. Today, the product of his labor is gathering dust in a basement room of Rockefeller Hall, arranged in pile after neat pile, chronologically ordered on shelves, one REVIEW number per pile. Of some issues (Number 4 of each volume particularly, for some reason) in Volumes I, II, and III, there are originals in the holdings, which shed light on covers and advertising. In the Cornell Library, Series I of the journal is bound without covers or advertising, and the first five volumes are of apparent modern, trimmed reproductions, with but an occasional original page, i.e., Marston's color print in Volume I is present therein. In Series II, some advertising pages are bound along with the science. Other libraries may have handled it differently.

Previous to the PHYSICAL REVIEW, there was no American journal devoted solely to physics. And why should there have been? Kévles writes that at the time there were only about 200 people in the nation participating in what they would call "physics," albeit most of that not very profound or basic in nature; largely it was applied, "practical" physics. The codification between the "pure" or abstract and the "practical" or applied has been with us from the beginning. The first American physicist of world note, Benjamin Franklin, was very much into applied physics: his spectacles, stove, glass harmonica, electrostatic motor, and such; but even he got into worrying about such basic problems as the nature of electricity, the thickness of oil films on water, and matters of meteorology and oceanography. We should perhaps not ignore among earliest American scientists, another Benjamin, Benjamin Thompson, born in Woburn, Mass. in 1753. He married a wealthy widow of Concord (then Rumford), N.H., where they lived for awhile. A loyalist and British spy in our Revolution, he fled to England, becoming a British citizen and Count Rumford. He had a checkered, many faceted and colorful career, doing his important science abroad, work which included, beyond fireplace development and the invention of a double boiler and a drip coffee maker, his elucidation of the nature of heat, and the founding, with Sir Joseph Banks, of the Royal Institution. He recruited for it Thomas Young and Humphrey Davy, the latter in turn hiring one Michael Faraday. Another, certainly American, was Joseph Henry, who discovered electromagnetic induction independently of Faraday, "pure" enough physics but Henry was nevertheless not above inventing telegraph signaling and telegraphic relaying.

After leaving Albany, where he made his discoveries, he went to a Professorship at Princeton and subsequently on to head the new Smithsonian Institution, leaving a fair imprint on the future of American science, even after his death in 1878. He was very much involved with the founding of the American Association for the Advancement of Science, Section B of which emphasized physics in its considerations, meeting even until recent times in its annual sessions jointly with the American Physical Society. He was a founder of the Washington Philosophical Society. As Secretary of the Smithsonian he was influential in arousing various fields of science and stirring them into activity; government support of science was encouraged, given some direction and purpose. He had no use for amateurism in science, or for those who could in a parlor lecture show a nice fossil or do a cute experiment, publishing their results as science in local "science" pamphlets. Kevles has him railing against such scientists, "puffs of quackery who can exhibit a few experiments to a class of young ladies," urging real scientists to get together and raise standards.

At the end of the century, in the decade of the nineties and of the REVIEW's founding, the truly great names in American physics were three: A. A. Michelson, who had already developed his interferometer and done with it the great fundamental experiment with Morley, a chemist; Henry Rowland, who was a strong advocate of the "pure" but not above getting into such applied matters as ruling excellent gratings, albeit as a means for investigating spectra; and, last but not least, J. Willard Gibbs, who, working alone at Yale, was certainly into "pure" physics, but whose work would nonetheless, in the hands of others, be of great practical importance. Thus should it be.

Rowland, in "A Plea for Pure Science," delivered to the A.A.A.S. at its Minneapolis meeting in 1883, worries about the pure *versus* the practical, and the paucity of significant physics being done:

> *"fain would I recount to you the progress in this subject (physics)
> by my countrymen, and their noble efforts to understand the order
> of the universe. But I go out to gather the grain ripe to the har-
> vest, and I find only tares. Here and there a noble head of grain
> rises above the weeds; but so few are they, that I think the majority
> of my countrymen know them not, but think that they have a
> waving harvest, while it is only weeds after all. American science
> is a thing of the future, and not of the present or past; and the
> proper course of one in my position is to consider what must be
> done to create a science of physics in this country, rather than to*

call telegraphs, electric lights, and such conveniences, by the name of science. I do not wish to underrate the value of these things, the progress of the world depends on them, and he is to be honored who cultivates them successfully. So also the cook who invents a new and palatable dish for the table benefits the world to a certain degree; yet we do not dignify him by the name of a chemist. (Seems to smack of snobbery.) And yet it is not an uncommon thing, especially in American newspapers, to have the applications of science confounded with pure science; and some obscure American who steals the ideas of some great mind of the past, and enriches himself by the application of the same to domestic uses is often lauded above the great originator of the idea, who might have worked out hundreds of such applications had his mind possessed the necessary element of vulgarity."

Strong stuff; the entirety makes for quite a speech.

He was right of course, that the public was more familiar with the names of such as Alexander Bell and Thomas Edison than that of Henry Rowland. Bell was pretty much a "practical"physicist but he was instrumental in helping support Michelson's experiment and was much involved with the temporary regeneration of the magazine *Science*, following some rough, discouraging times it had experienced (and would still experience). Edison, on the other hand, was nothing but an "applied" physicist, having complete disdain for abstract and theoretical physics; "physicist" is perhaps going too far in labeling his applied vocation. Kevles quotes him: "Oh these mathematicians make me tired. When you ask them to work out a sum, they take a piece of paper and cover it with rows of A's, B's and C's, scatter a mess of fly specks over them and give you an answer that's all wrong." Yet he was not above having a mathematician on his World War I Naval Consulting Board, realizing, as he was reported to have said in connection with its constituents, that "very few really practical men are expert mathematicians...it is advisable to have one or two men...who can figure to the nth power if required." n^{th} power? Edison? Or his spokesman (Chief Engineer, H. R. Hutchison) putting words in his mouth? Edison's figure loomed large but rather ineffective in that war's science effort.

And where did such men publish? In Franklin's day there were no American journals; his findings went across the Atlantic in letters to such as the Royal Society of London, the *Philosophical Transactions* of which, founded in 1665, is today the oldest surviving scientific journal. Henry published in *Silliman's Journal*, which became the reputable *American Journal of Science*, the latter carrying his discovery of mutual induction (which Henry somehow delayed in submitting, so giving the priority to Faraday). Rowland submitted one of his first investigations to the *Journal* but in the hands of "referees" at New Haven, where it was published, he was rebuffed, which prompted his sending the paper directly to Clerk Maxwell, who quickly had it published in the *Philosophical Magazine*, much to Rowland's satisfaction and reputation. Nonetheless, a number of Rowland's collected (some sixty) papers were first published in the *American Journal of Science*, which journal, much to its credit, in November 1887, Volume XXXIV (3rd Series) published a paper, Art. XXXVI, "On the Relative Motion of the Earth and the Luminiferous Ether" by A. A. Michelson and E. W. Morley, following by six years an article in the same journal by Michelson from the Naval Academy on his first attempt at the measurement, in Berlin and Potsdam, with his new interferometer, there described. In that paper he thanks Mr. A. Graham Bell, who has provided the means for carrying out this work. Michelson more often published in European journals, and poor Gibbs relied on the obscure *Transactions of the Connecticut Academy of Arts and Sciences*. Edison, of course, "published" extensively in the Patent Office and his contributions were eagerly fed to, and reported in, the daily press and popular science magazines such as the *Scientific American*. Bell, considerably more enlightened in science than Edison, also got patents, one particularly very well known and significant; but he also delved in more abstract matters. (One must be fair in this, however much we fault Edison for his view of fundamental sciences; his contributions certainly greatly impacted society. And, while he made nothing of it, he *did* discover one important fundamental phenomenon—that of thermionic emission in what was called the Edison effect, a curious effect in his lamps, nothing for him to worry over, understand, or investigate.) Bell, with his father-in-law, Gardiner Hubbard (both also to be involved in what is today's popular monthly, somewhat misnamed *The National Geographic*) gave support and new life to the failing magazine *Science*, a mix of pure and applied sciences in various fields. The cure did not take; the magazine folded a decade later to be bought from Bell by James Cattell, who in 1900 also took on the editorship of *Popular Science Monthly*, founded in 1872 and influential in spreading scientific thought. It came into being to "create a scientific culture" as Kevles puts it. In both journals Cattell was outspoken in support of science, general interest in which was finally beginning to stir. Tyndall had made a very successful and widely reported lecture tour of the country, exciting much interest. Huxley did likewise, lecturing on Darwin's great and, to many, unsettling proposition, during the tour giving the inaugural address of Hopkins' first president, Daniel Gilman, who had given up on Berkeley for this new, promising institution in Baltimore. Abroad, Helmholtz was expounding on Joule's concept of heat, extending the view to the all-important notion of the conservation of energy, duly reported in the popular American journals.

One of the early popular journals, of importance not to be discounted, and still with us, was *Scientific American*. It was founded in 1845 as a weekly patent journal, the publishers actually coming to serve as something of a patent agency and referral service. Inventing was going on at a lively clip. The weekly gradually began reporting also on news of astronomy and medicine, and increasingly informing an interested populace on science generally. It became a monthly, with less on patents, in the early nineteen-twenties, publishing readable and authoritative articles of general interest across all of science. While not a research journal at all, it is today widely read and highly useful.

Thus in the latter part of the century, middle class Americans were being exposed to and stimulated by the new disciplines. Many new colleges were springing up with chemistry and physics emphasized. It was, however, in the leading private institutions where the science would be abstract, largely "to produce the well balanced and liberally educated man," to quote Kevles. Cornell's Anthony, predecessor of Nichols, and a practical physicist as we will see, saw the problem thus: "In this country, men devoted to science purely for the sake of science, are and must be few in number. Few can devote their lives to work that promises no return except the satisfaction of adding to the sum of human knowledge. Very few have the means and inclination to do this."

So the time was ripening for the appearance in this country of serious journals in science. The earliest American journal of note was the above mentioned *American Journal of Science*, published at Yale, "James D. and E. S. Dana, Proprietors." It was founded in 1818 and was devoted largely to the work of amateurs until, under the prodding of Joseph Henry, it became a professional journal containing a mix of mineralogy, fossils, geology, biology, and the like, and such physics as there was. Kevles reports that in the last quarter of the century it had no more subscribers than it had at the beginning and was losing money. The journal was started by Benjamin Silliman, Professor of Chemistry and Mineralogy at the New Haven institution, and was frequently referred to as *Silliman's Journal*. James Dana had been an undergraduate student in Chemistry under Silliman and later laboratory assistant to the Professor. After four years as geologist and mineralogist on the U.S. Wilkes Pacific Expedition, Dana returned to Yale, married Silliman's daughter and became, at the old man's retirement, Professor of Natural History and Geology, meanwhile associating himself with the *American Journal of Science (and Letters)* as associate editor with Silliman. Dana's son, Edward S., rather followed his father, also becoming a Yale professor (in natural philosophy) and, with James D., joint "proprietor of the *Journal*." It is little wonder that the magazine devoted itself rather heavily to geology and minerals. Kevles writes that Dana had recourse only to rather mediocre physicists for judgments on those physics papers that did come in, which is probably why Rowland received his brushoff. Further: "Dana evidently tended to treat generously articles submitted by physicists better established and less mathematical than the young Rowland. Compared with the number of papers in geology, few papers in physics came along."

Another early journal was that of the Franklin Institute, founded in 1826 at the Institute, which was established a few years previously in honor of the great man. Papers were pretty much devoted to practical ends. Rowland made some use of this forum. Late in the century, there were of course well established physics journals in Europe: The *Philosophical Magazine* and the *Proceedings of the Royal Society*, both going way back; there was *Nature, Comptes Rendus, Annalen der Physik, Nuovo Cimento*, and so on. It was thus, in 1893, a brave Cornell crew who hoped in their PHYSICAL REVIEW to emulate in this country such prestigious publications. But with the scientific ferment abroad the land, and interest in things physical, the time was suitable for a true American journal of physics, even if the American physics being done was largely mundane and dull, however laboriously carried on. Actually, Rowland, at Hopkins, had proposed nine years earlier, in 1884, that his institution sponsor a physics journal. But Dana, at Yale, willing enough to kill the project, gave assurance to D. C. Gilman, Hopkins president, that the Dana Journal, *The American Journal of Science*, would "give early publication of any physical paper that may be received"; of interest to Rowland no doubt.

Prestige was a long time in coming to the PHYSICAL REVIEW; European physicists in some quarters considered it something of a joke. Professor Rabi tells in John S. Rigden's biography, "Rabi," of going in 1927 to the physics library at Hamburg University on his first trip as a physicist to Europe. He sought the REVIEW to learn what was going on at home and was shocked at finding no current copies on the shelf. To save money, the library got a year's worth of issues all in one bundle. It was not worth the extra postage to get the journal on a regular monthly basis. But prestige did come. "Ten years later," Rigden notes, "the PHYSICAL REVIEW was the leading physics journal in the world." Kevles has de Broglie agreeing, saying in the mid-thirties: "Today scientific publications from the United States are awaited with an impatience and curiosity inspired by no other country."

Today, the PHYSICAL REVIEW appears in various sub-fields of physics, each issued biweekly or monthly, where initially one "flavor," encompassing all, came every couple of months. Today some single issues embody more pages than a year's six issues back then. For more rapid publication of "important" work, we have the weekly PHYSICAL REVIEW *Letters*, sprung from the REVIEW and rather more widely read than its enormously more voluminous parent, which nobody can possibly read (or want to) in entirety. Additionally, there is the *Reviews of Modern Physics*, offering quarterly reviews of this or that field; and also the biweekly PHYSICAL REVIEW *Abstracts* "to provide the physics community with advance information about work to be published and thus give early notice to researchers of related work done by others." Rather easier to scan than the complete REVIEW, the *Letters*, and *Reviews of Modern Physics*. The *Journal of Applied Physics* prints monthly what its name implies; the *Review of Scientific Instruments* monthly relieves the PHYSICAL REVIEW of publishing papers on new instrumentation, which were a staple of the early REVIEW. The journals are all served today by the American Institute of Physics, which also has under its wing several other specialized journals covering fields such as optics, acoustics, rheology, and the like. The non-specialized *Physics Today* brings news of physics, of physicists, summaries of new discoveries, book reviews, a calendar of events, obituaries—topics of general physics interest, some of which the early PHYSICAL REVIEW tried to cover. Abstracts of meetings of the American Physical Society used to be given in the REVIEW. No longer. Abstracts of papers of some meetings today number in the few thousands and may take up five or six hundred pages of the *Bulletin of the American Physical Society*. In 1993, for example, the Society sponsored 40 meetings, General, Divisional, Sectional, and Topical (conferences), in which a total of at least 12,000 contributed and invited papers were given! The largest meeting by far of any year these days is the March meeting devoted mostly to solid state physics, in which there were about 4000 contributed papers and invited talks, with 4500 physicists in attendance.

At this point it is appropriate to try making clear the part that the American Institute of Physics has played in the publication of the PHYSICAL REVIEW. There has been, and still is, considerable uncertainty in distinguishing the role of the American Physical Society from that of the Institute in the publishing of our journals; a lot of trouble to the Society has ensued in consequence. Until the establishment of the Institute in 1932 there was no problem, there was only the Society and it was publishing Series II of the PHYSICAL REVIEW, clear and simple. From 1933 until 1980 the cover page of the PHYSICAL REVIEW indicated that the journal was "Published for the American Physical Society by the AIP." In the fifties, about the time Goudsmit became Editor, and despite the cover statement, the Society is said to have assumed the greater part in the publishing enterprise ("the actual publisher") but it was not until 1980 that the REVIEW cover reflected this fact, in notice that it was published *by* the Society *through* the Institute. So it is now also with the other journals of the Society: PHYSICAL REVIEW *Abstracts*, *Reviews of Modern Physics*, and the *Bulletin of the American Physical Society*. PHYSICAL REVIEW *Letters* is the Society's alone, and always has been. The Institute has the role of performing for some of the publications certain paid service functions. It takes care of the subscriptions; for the PHYSICAL REVIEW and the *Reviews of Modern Physics*, it handles the editorial mechanics and composition; and, except for PHYSICAL REVIEW *Letters*, it does the subcontracting with Lancaster Press for the printing and mailing. The Society owns the journals, sets prices and editorial policies, and completely manages all aspects of their operations. The Institute does the same for its own journals; and it also performs similar services for journals put out by other member societies, publishes English translations of a number of Soviet journals, plus books and conference proceedings. The Institute has no individual members of its own, but is rather a consortium of its member societies, the largest of which is the American Physical Society. End of clarification.

American physics (as is true of world physics) has indeed become a very large enterprise. The glut of information pouring into publications' offices is becoming unmanageable, if it has not already done so. In 1962, in discussing some major technical problems of the coming next twenty-five years, Dale Corson, then Cornell's Dean of Engineering, considered the handling of information. The time period is, at this writing, just up and his concerns are no less today than they were then. He relied on figures given by Yale historian, Derek Price (see Price's book, *Science Since Babylon*). From about the mid-1700's when there were ten or so journals, the growth in number of technical publications up through 1950 was remarkably exponential in character, with a doubling time of about 15 years, a factor of ten every fifty years, so that by 1950 there were about 100,000 technical journals in the world. From that point on, the doubling time seems to have sharply decreased to about 8 1/2 years. The number depends of course on what is taken to be a technical journal; presumably the *Christian Science Monitor* and the *Herbach-Rademan Monthly Electronics Catalogue* do not count, where the *General Radio Experimenter* and the *Hewlett-Packard Newsletter* may well be included. (Price worries about the weight in the count of the PHYSICAL REVIEW as compared to the *Annual Broadsheet of the Society of Leather Tanners of Bucharest*!) As Corson says, however, the actual number is not too important, for even if one took the number to be only (!) 50,000, in 8 1/2 years it would be 100,000. By 1830, when there were some 300 journals, consternation had already set in and the first journal of abstracts came on line. Growth in the number of these has been the same; in 1950 there were some 300. If the trend continues, the year 2000 should see the first journal abstracting journals of abstracts! It is said that in 1962 on the order of a thousand pages of text were being printed every minute of the day, seven days a week, fifty-two weeks a year; that's a twenty-four hour day—Christmas no holiday. A small army is required to crank the stuff out. Incredible, if even half true. Today there must be on the order of a million technical journals of one sort or another! We are inundated, engulfed. Physics as a discipline, and in particular PHYSICAL REVIEW, has partaken of this same growth. Professor Mermin, at Cornell, has found the doubling time in yearly REVIEW pages to be about ten years since 1936. In the Clark Physical Sciences Library at Cornell, for striking illustration, the complete bound volumes of the REVIEW up to the year 1926 when John Tate took over the managing editorship (journal age 35) occupy about nine feet of shelf space. In fifteen more years, at the beginning of World War II, this had approximately doubled to nearly eighteen feet. By 1994, over 200 feet (!) of shelf space has been taken over by the journal; and that does *not* include about 30 feet of PHYSICAL REVIEW *Letters* and twelve feet of *Bulletins of the American*

Physical Society, both of which used to be an important part of the REVIEW. These figures are not quite double those of ten years earlier, so perhaps the rate of growth estimated by Mermin is slowing down. In 1926 there were a total of 1850 members of the Physical Society, up from 495 in 1909; in 1946 there were 5700 members. Today we number over 40,000. Little wonder that the REVIEW and its cohorts have literally exploded. David Lazarus, then Editor-in-Chief of the Physical Society, in his May 1991 *Physics Today* appreciation to W. W. Havens, outgoing Society secretary, stated that the Society then had a full time staff of over 140, published some 70,000 pages "of magnificent physics," and sponsored 45 national and internatioal meetings and conferences each year. The annual budget was $20 million. Not only has the number of physicists increased tremendously, along with their publications, but so also has the magnitude of their researches, in many cases a single project (and publication) involving a multitude of people (and coauthors) and expenditures in the millions of dollars (if not billions!). In his worrisome discussion Corson alluded to the growing role computers would have in the storing and retrieval of information; much of his forecast has been borne out, and further progress is being made in an attempt at weathering the threat. But will it be able to keep ahead of the storm?

It is interesting, but perhaps not suprising, that the REVIEW's growth has paralleled that of modern physics itself: Roentgen discovered X-rays in 1895, two years after Volume I, Number One, and Becquerel's discovery of radioactivity came a year later in 1896. Zeeman found his effect the same year, and Thompson discovered the electron in 1897. Following them in orderly succession came Planck (1900), Einstein (1905), and Rutherford (1911), culminating in Bohr (1913). Then after a period of murk came the clearing provided in the mid-twenties by Pauli, de Broglie, Schrödinger, Heisenberg, and Dirac. Altogether an amazing thirty years progress. Yet the PHYSICAL REVIEW chronicled or got involved with very little of this in the same time period—its own first thirty years. The foundations and development of modern physics were laid out in Europe; European publications carried the revelations and most of the experimental support. American physicists trooped to Europe to learn and engage in the activity first hand. It would be perhaps forty years after the birth of the REVIEW before this situation would dramatically change; as alluded to earlier, European physicists would be coming to America, the journal would be first to describe many discoveries in physics worthy of, and recognized by Nobel prizes, and be publishing papers as deep and as seemingly fanciful to many as did various of the European advances in physics seem to most American physicists in the first decades of this century. How much the growth of American physics may be attributed to the REVIEW is unanswerable; perhaps the growth of the journal was a result of the growth in American (and world) physics. They certainly have come along together—both large enterprises today, each fostering the other.

It is the purpose of this review, celebratory but unscholarly, to trace the founding and development of the PHYSICAL REVIEW, especially in its relationship to Cornell, and to give some indication of the sort of journal it has been, mainly over the first fifty years, and to mark its gradual maturation into what is almost certainly the world's leading journal(s) of physics. The world is a very different world than it was one hundred years ago. It is true, if trite, to say that we have come from the horse and buggy (with its pollution) to the gasoline engine (with its own pollution), from kite flying to the jet and man on the moon, from the telegraph to near instant global television. No less parallel may be drawn regarding the PHYSICAL REVIEW.

To give some feeling for the fifty years' development of the journal from its primitive beginnings, we propose noting significant events in the journal's course, and picking out in some sort of chronological order papers that are typical (we hope), landmark papers, curious papers, so-so papers by well known physicists, and papers which also mark the growth of physics generally during the period. Not only will we cite authors and their articles; three years after the founding of the American Physical Society, an event coming five years after the REVIEW first appeared, the Proceedings of all Society meetings (save the first seventeen, which were printed in the short lived early *Bulletin*) were an irregular feature of the journal. As such, we make reference to many Proceedings of meetings of the Society, in a way thereby also sketching somewhat the development of the American Physical Society, with which the PHYSICAL REVIEW can hardly be dissociated. In addition, we will cite a great many abstracts (or titles only) of papers presented at meetings, which may or may not have been followed by full-length articles in a subsequent issue of the journal, or may have been sent for publication to another journal, most frequently to the British *Philosophical Magazine*.

The early REVIEW did not exactly take prompt notice of developments in the field from abroad, which were greatly influencing modern physics. We will try, therefore, to tie in landmark European work with what was going on over here and in the journal by citing some pertinent items from *Science Abstracts*, which for a very long time came along with one's REVIEW subscription as a member of the

Physical Society; we connect the two journals during the period as going hand in glove together.

Whether the overall attempt we make will be at all successful will have to await the judgment of the reader. It is not clear that in a "history" of a technical journal, anything of general interest can result; we are not a writer for the *New Yorker* putting together a work about that popular magazine; there, many in the cast of characters are still around. For the PHYSICAL REVIEW none are. What can be said, however, is that it has been an interesting exercise for the writer to have gone through fifty years of the REVIEW, in a way to have met many well known physicists, to have experienced their discoveries, and the developments they have advanced, and to have watched "first hand" in a way, the growth of our science during the exciting period which this survey tries to encompass. The experience is one that anybody with a complete set of the journal volumes may have, given some time and the incentive; so no scholarship, great imagination, or originality will be found herein. But it may possibly be of interest to some and serve the purpose of celebrating the journal's centenary.

The citation of paper after paper may prove boring, especially in the referencing. Volume number and dates of cited articles seemingly must be given, and at least somewhat chronologically, and for the most part be included directly in the text, not as footnotes. In other instances the referencing can occur parenthetically, especially in the brief survey of post World War II developments. Over much of the latter period, PHYSICAL REVIEW *Letters* is also involved so that the particular journal must be specified for that period. In attempt to avoid tiresome repetition in the manner of referencing, the mode will vary, abbreviation may appear. Paper titles will not always be spelled out and page referencing seems not necessary; this is not a scientific treatise in the usual sense.

Issues of the journal were numbered in Roman numerals consecutively from I in Volume I (July–August, 1893) through CXCIX (deciphering is left to the reader) in Volume XXXV (December 1912). The new Series II, starting in 1913, no longer continued the sequential numbering of issues but stuck with the Roman numerals in the Volume identification until a new editor and editorial council came aboard in 1923. Abruptly, Arabic notation was recognized. We will follow suit herein with the Volume labels.

Chapter 1
THE CORNELL SCENE

•◆•

At the time of the first appearance of the PHYSICAL REVIEW, Cornell University was but twenty-five years old; the first students were admitted in 1868. The Physics Department was of like vintage; it had been present at the beginning, which, in spite of there being at the time only about 75 people in the country calling themselves physicists (Kevles), is not too surprising. The University was founded by Ezra Cornell, who had come into his wealth through his association with what was to become the Western Union Telegraph Company, made possible by electromagnets "discovered" by Arago, Davy, and others in the 1820's. Ezra Cornell invented a plow which could both dig and cover a ditch holding a buried cable. Actually, because of insulation difficulties with cable of the time, a line was instead strung adequately on poles between Washington and Baltimore to carry the famous message, "What hath God wrought?" For all his rough cut, down to earth, rural character, Ezra was a somewhat impractical sort. He wanted a university where anyone could study any subject, even though it be shoe cobblering. He wanted the world's largest telescope in Ithaca; skies were perhaps more transparent in those days than they are today—on the occasional really clear night they were surely darker. It took some forbearance on the part of his friend and erudite colleague, Cornell's first president, Andrew D. White, to steer him away from such ventures. White was almost the complete opposite of Ezra Cornell; an educated, cultured gentleman, a classicist. He was a scholar like Gilman of Hopkins and knew the value of "pure" science. It is curious that in spite of White's earlier experience in Germany, where the university as a research center was well established, he only speaks of "scientific study" in his inauguration address, making no mention of research, which today is as important as teaching in the university mission. But research has been a large aspect of Cornell University's Physics Department from the start.

In spite of his munificence, Mr. Cornell must have been a trial to the fledgling institution. He was ever present, protective and intrusive, which was probably in some measure responsible for the departure of the first Professor of Physics, Eli Blake, grand nephew of Eli Whitney of cotton gin fame; he left before two years were up to take a position of long tenure at Brown. In a visit over at Troy University with one Professor Rood (whose sister was Blake's first wife and who himself wrote a technical treatise on color which became something of a "bible" for pointilist artists) while en route to his Providence post, he apparently unloaded his frustration. Rood wrote Columbia's Professor Rutherfurd (for whom the Rutherfurd observatory atop Pupin Hall is presumably named) about that visit. Mr. Cornell was into everything; he wanted Blake to set up shop practice for his boys in the Department. Blake demurred, saying it was not exactly physics and it would take too much time. Cornell countered with the remark that he had learned in two weeks all that he knew of tools; "enough ignorance and stupid ideas to overstock six ordinary men," was Rood's comment to Rutherfurd. But Blake also had trouble with White. For start up, the Department had been allotted a full $3000 for apparatus, which Blake assumed would be his wisely to spend. But no. White was soon off on a buying trip to Europe and spent practically the whole sum on a fine projector without batteries, consequently unusable. For a small speculum mirror that Blake suggested be acquired, White brought home a lump of speculum metal acquired at an apothecary's shop.

So Blake left Cornell. Unfortunate, but it made possible the appointment of the man who really set the course of the Department, and was eventually to name his own successor in the job, the man who would found the PHYSICAL REVIEW. Blake's replacement was William Anthony from Iowa. Anthony was a sensation from the start, a strong believer in laboratory work, and in the demonstration lecture. Under him, Cornell was one of the few institutions offering laboratory work in physics. He gave popular public lectures, charging admission to make possible the acquisition of apparatus (beyond a fine projector). While he brooked no foolishness, he must have pleased Mr. Cornell with his practicality. With a former student, George Moler, Anthony built what was long regarded as the first practical electrical generator in America. There seems to be some question about priority in that claim but it was a reasonably large machine (still extant), self excited, steam driven, used in the "first" lighting system in the country (at least on a university campus) powering as it did a sputtering arc atop a college tower, farmers across the lake valley knowing in the night that the institution was still there. Electricity was *the* research area in Cornell physics. Anthony built a giant, multi-coil, tangent galvanometer (an instrument no longer treated in elementary physics text books) capable of measuring currents from milliamperes to a few hundred amperes. Electricity was the "in" discipline and remained so, in spite of Harvard's Professor Lovering's labeling the interest as "only a spurt." Indeed, the field of Electrical Engineering grew out of the Department's electrical activities and was a side arm of the Department for years. (The same happened at MIT. Robert Rosenberg in an—October 1983—*Physics Today* article has related the circumstances connected with the two developments.)

The successes and progress in electricity were not lost to the public eye, only enhancing further progress in science generally. The impact of the 1876 Centennial Exposition in Philadelphia was also noteworthy. There high technology was exhibited, "oh-ed and ah-ed": Edison's lamp, Bell's telephone, Anthony's generator, a huge Corliss steam engine, all functioning, all stimulating and exciting to the populace seeing them or reading of them. The role of the exposition in the rapid growth of science was not negligible, as Professor Bedell wrote in recollection [*Phys. Rev.* **75**, 1601 (1949)].

Matters electrical remained a strong interest of the Department all through Anthony's fifteen years. He supervised the installation of the Ithaca R.R., a short electrical trolley that ran cars up State Street from the Lehigh Valley R.R. station to the Ithaca Hotel uptown. The system later expanded so street car transportation was available to the University and hillside community for many years. In fact, Ithaca R.R.'s first generator subsequently came to the Physics Department and served out its years as the magnet supply for the small but notable cyclotron built in the mid-nineteen thirties. The Department's dynamo laboratory was a continuing outgrowth of this interest in electrical machinery.

President White resigned his position in 1885, having guided the University over its formative first twenty years. He was succeeded by his protege C. K. Adams, who had little use for scientific study, and was distressed by what he saw as Cornell's lack of scholarly learning. The time for the classics had come. Anthony and Adams just could not see eye to eye. In contrast to Anthony, Adams was somewhat secretive in his dealings. Rather than consult his friends like White, Anthony simply resigned, leaving the University for "industry," and taking a position with Mather Electrical Company in Manchester, Connecticut. He went back to teaching in 1894 at

Cooper Union, where he remained until his death in 1908. As his Cornell replacement he recommended a former student, Edward L. Nichols, telling friends that he thought the Physics Department could go much further under Nichols than it ever could under himself. By that time Nichols had had four years of graduate study and experience in Germany behind him, had returned to this country to work with Rowland at John Hopkins and with Edison in New Jersey, had taught in Kentucky for a couple of years and at the University of Kansas for four; and he was a natural to assume the Department leadership.

Chapter 2
A New Journal
Volume One

◆◆

In 1892, Cornell selected as its new President, J. Gould Schurman. This was the year at the institution that Ernest Merritt became an Assistant Professor and Frederick Bedell earned his Ph.D. And it was the year apparently when the notion of starting a journal devoted solely to physics came to the mind of, and was put into action by, Edward Nichols. It may well have been that the seed was actually planted by the new president; he was himself the editor of two journals: the distinguished *Philosophical Review*, which was started at Cornell that same year (first issue: Jan. 1892) with Schurman as editor, and the influential *School Review*, which he also established from Philosophy (Education and Pedagogy being part of Philosophy in those days). Schurman encouraged his faculty likewise to take on such obligations. If, indeed, the seed was planted after his accession to office, not much time was wasted in germination; the first issue of the PHYSICAL REVIEW came out in the summer of the following year, "Conducted by Edward L. Nichols and Ernest Merritt," as we have noted.

The desirability of there being an American journal dedicated to physics had certainly occurred to others and likely to Nichols before the advent of J. Gould Schurman. D. C. Gilman, Hopkins' president, had already rejected Rowland's suggestion for such a journal to be subsidized by his institution. But the encouragement of Cornell's president would certainly have added incentive to Nichols if he had had it in mind.

The proposal to found an American journal of physics was an audacious undertaking. Not only were there the reputable and established physics journals in Europe but there were not many physicists working in this country and, while they might not have recognized it, much of what they were doing was not very significant physics. And there was already the *American Journal of Science*, which published physics related papers, tucked in amongst the

20

mineralogy, fossils, and geology. What passed for physics was surely related enough to physics to merit inclusion in a journal devoted to physics, but it remained largely "fact gathering," as Kevles labels it. Rowland, Michelson, and Gibbs were our exceptions, not discounting Bell and Edison, known best to the public but primarily inventors, a breed somewhat looked down upon by so-called "physicists." On the other hand, "Fact gathering" should not be entirely belittled; must it not precede synthesis, theoretical understanding, and the great "breakthroughs"? In spite of this somewhat disparaging view of American physics being done at the time, one must not lose sight of the few real contributions; in particular, perhaps the greatest experiment of "modern" times had already been done, and done in this country. In 1887 at what is now Case Western University, Michelson and Morley (of Chemistry) performed their great opus. While very important to the support (apparently not the *raison d'être*) of Einstein's special theory of relativity coming twenty years later, the experiment stood disturbingly rather alone, mysterious and essentially apart from the rest of physics developing around the constitution of matter.

The whole business of organizing the launch of a new publication is not something lightly to be undertaken and accomplished in a weekend or so. Listing of possible subscribers and solicitation for support and contributions have to be generated and circulated, notice of the proposal has to be advertised somehow, advertisers themselves solicited, printing and mailing to be arranged and, most important, financial backing for the enterprise to be established, not to mention that enough contributed papers be in hand and guaranteed to be forthcoming. Toward this last, Cornell came through. Physics staff, past and "present" associates contributed heavily in articles, and the University administration notably undertook the monetary support, undoubtedly with the enthusiastic concurrence of President Schurman. A memo dated May 3, 1893, found in those "archives" of the Rockefeller attic, goes simply:

To: Professor E. L. Nichols
     Ithaca, New York

The Executive Committee yesterday appropriated $500 for the PHYSICAL REVIEW

Yours truly,

(signed with whorls and great flourish)
E. L. Williams

There were probably many other such memos from Williams; Bedell indicates the first year's appropriation(s) came to $2400. Macmillan was contracted to do the publishing. Subscribers were garnered through letters, notices given in other journals, and the like. Together with their teaching obligations, this whole endeavor must have kept Nichols and Merritt rather busy, and staff working at the preparation of articles for publication. Cornell Professors Howe and Grantham, later in an abbreviated history of the University's Physics Department, note the following as regards the founding of the REVIEW:

> *Professor Nichols was hardly settled at Cornell when, concerned over the lack of an American journal devoted exclusively to physics, he persuaded the University to furnish financial backing for such a publication, to be edited by him and Professor Merritt. Thus came into being the PHYSICAL REVIEW, now so familiar to all American physicists. Bedell, among the first group to receive the Ph.D. degree in physics at Cornell (1892) was added to the editorial staff before the first copy of the REVIEW appeared in 1893. For twenty years the three men managed and edited the journal. When this Nichols–Merritt–Bedell–Cornell enterprise became self-supporting, it was turned over, outright, in 1913 to the American Physical Society. After that, Bedell served for still another ten years as managing editor, from his office in Rockefeller Hall.*

Response to the announcement of the proposed new journal was encouraging. One of the earliest was that from the first Cornell Physics Department head, Eli Blake, over at Brown. After some discussion on the silvering of mirrors, in his handwritten letter from Providence, he expresses interest in the undertaking—"the era of specialization is on us," he notes, thinking perhaps of the mixed bag of sciences to be found in the *American Journal of Science*. Benjamin Snow wrote from Indiana: "But few pieces of news have of late given me greater pleasure than that contained in your last letter in which you mention your plan for starting a journal devoted to physics. That there is room for such a journal we all know and I am more than glad that you are willing to take upon yourself the certainly no easy burden of carrying it on." He will be glad to help however he can. From Professor Ayers, Tulane Physics and Electrical Engineering, came the comment that Nichols was just the man "to carry off the enterprise." A handwritten letter from K. Angstrom in Sweden (son of *the* Angstrom, for whom we have named a unit of wavelength) to Nichols accepts a request to publish (and translate) a paper in "the journal about to start"—he wishes every success. One from MacFarlane at Texas; he is pleased to learn of the new journal—he offers to review books on mathematical physics. Later, in another communication, he's reading Heaviside's Electrical Papers and hopes for "not a mere description of the books but a critical notice, especially of his system of vector analysis." Praise for the idea came from Henry Crew (Chicago), and from "James D. and E. S. Dana, Proprietors" at Yale publishing the *American Journal of Science*. Crew wrote that there would be a "certain comfort in not having to look for American work sandwiched in between Dana's minerals and Marsh's fossils. E. S. Dana himself wishes the journal "all possible success ... count upon our cordial support and sympathy in the undertaking ... it seems as if the country were large enough to support it well ... ." (Relations apparently were cordial; a few years later Bedell wishes to exchange journals and invites Dana up to Ithaca for a visit. Dana sends two copies of his *Journal* but unhappily cannot accept the invitation: "... while not relegated to a sanitarium ... am still as near being a cipher in the community as an unfortunate individual can be." A slight exaggeration; he was with us until 1935.) Samuel Sheldon from Brooklyn Polytechnical Institute has an article for the *American Journal of Science* but will put it in the REVIEW if they'd like it; he doesn't much "like physics buried in a journal, three fourths of which is mineralogy and geology." He would come to have a paper in both Volumes II and III of the REVIEW. So with Rogers at Waterville, Maine (Colby College); he too has an article (on the yard *versus* the meter) ready for the Dana journal but would prefer to have it in the PHYSICAL REVIEW if suitable. It was, appearing in the introductory issue. The journal "will fill the bill" to get the results "from your laboratory," wrote S. E. Hill from Rockford College in Illinois. Our journal, parochial enough, was not intended to be that parochial.

The name of the proposed journal engendered some concern. From Haverford, J. O. Thompson wrote Nichols that he was glad to hear of the journal; his old Professor Kohlrausch urged students to put physics papers in one quarter so as not to be scattered all over the place—"wonders about the name"—"is PHYSICAL REVIEW exactly appropriate?" More or less to the point also was this from T. C. Mendenhall of the U.S. Coast and Geodetic Survey: he was just back from a "sort of British lion tail twisting trip to Montreal." He thinks he "gave the aforementioned appendage an additional torque (!?) of half a turn or so"; no reason given for the animal abuse but he hopes for acceptance of apology for not writing sooner, about the notice of the forthcoming publication. He's in favor of the journal but worries about the name; PHYSICAL sounds too gymnastic and out-of-doors. He suggests "Journal of Physics" or "Review of Physics," etc. Of course, any name with the word Review in it opens the door to punsters. Cornell is certainly not the only institution to have offered questionable entertainment in a Physical Revue ("Better Jazz through Science") nor the only abode of physicists who, harrassed by their accumulating journals, have disrespectfully referred to their holdings as "physical refuse." We leave unsaid what can be made out of the unavoidable "physical."

We may have come close to not having our name. The letter from Knut Angstrom is revealing. After thanking the editors for accepting his paper and expressing best wishes for the new journal, he indicates that his paper will "appear in *Wied. Ann.* one of these days." But he is glad to translate it for American and English readers—his wife is a big help. A better paper is one he has done earlier on the absorption of gases—in French. "When the German journal *Physikalische Revue* was started last year, it was with pleasure that I saw it overtake (undertake?) their translation, being thus saved that trouble myself. This journal existed however but a year and now my perhaps best paper is buried in the transactions of the Stockholm Acad. and in that journal that never obtained many readers." Etc. Interesting.

MacGregor from Dalhausie sent congratulations on the proposal: "...the study of physics has been advancing in America at such leaps and bounds in recent years ... there must be sufficient intellectual backing." Nichols had apparently suggested that MacGregor submit some work he'd done but MacGregor says he could not very well publish the same work in two places but he'd gladly submit future work to the REVIEW. (He did—one paper and one book review.) LeConte Stevens, over at Rensselaer, in a nice, thoughtful, encouraging letter, did not worry too much over the duplication problem; he had nothing to contribute himself at the moment but he would write E. H. Loomis that the latter's "Arbeit" with Kohlrausch should be submitted as a paper for U.S. readers. It was—in Volume I—"The Freezing Points of Dilute Solutions." At least the duplication in that case would be in two languages, but apparently it would not be uncommon that the same paper would go to both the REVIEW and an English journal, the *Philosophical Magazine* or *Nature*, for instance. A letter from Fernando Sanford of Leland Stanford Junior University says he will be glad to publish in the journal; he subsequently did, and frequently. Another letter of his accompanied one of his first submissions—on "electric photography"; pinned to the letter is a small faded print showing the features on the face of a silver dollar (silver was the metal in those days, and dollars came as coins—at least out West) registered by the phenomenon, one of two prints he had submitted with his paper.

In the discovered Rockefeller Hall attic repository were also many, many letters to and from Macmillan, the first dated February 8,

1893, in which Macmillan indicates they would be glad to publish the new journal. To Nichols they write: "... we beg to say that we shall be very happy if arrangements can be made to publish the journal for the University." But they need more information: "Our suggestion in regard to the matter, however, would be that the work should be undertaken at your cost and that a commission should be charged on subscriptions taken and copies of the periodical sold." "We furnish bills ... etc. ... for work done ... and other expenses of publication, etc." Details would have to be worked out. Such indeed followed: what size of page, weight of paper? trimmed edges or uncut? Apparently uncut was agreed upon; frequent pages in the Cornell Science Library holdings are *still* uncut! A number of years later a letter to Merritt from A. E. Kennelly objects to the uncut pages; he has calculated the total man hours for all subscribers spent in cutting them; the journal could avoid much profanity if the pages were cut and trimmed. Another letter supports him in the objection. But the REVIEW came uncut and untrimmed for many years to come, up to Volume XVII, beginning in July 1903. Actually, the editors had suggested to Macmillan, before the change was made, that they go to trimmed pages. Macmillan pointed out that it would increase paper costs (through wastage) by about 7%. That route had been taken by *Science*, and it was found that a lot of subscribers preferred (!) to trim their own. Macmillan could arrange, however, to supply trimmed pages to those requesting that format. Indeed, the inside of a front cover (at least that of Volume II, Number 4) notes that "Subscribers who desire to have their copies of the PHYSICAL REVIEW sent to them with the edges cut, can so obtain them by informing the publisher to that effect." The subscriber was on his own; Kennelly had just been asleep at the switch. After Volume V, the same note appears as an item in the information to authors and subscribers on the inside of front covers. There one also learns that "Authors of original articles published in the REVIEW will receive one hundred separate copies in covers for which no charge will be made; additional copies when ordered in advance, may be obtained at cost." In submitting articles "... when publication in other journals is contemplated, notice to that effect should be given." In May, before the first issue, the publisher worried about the price to the "Trade and subscription agents." No price had yet been given them; with a retail price of $3.00 (per year presumably), would $2.40 be okay? That would include postage.

In April, Macmillan was making progress: "... the enclosed proofs for the cover for the PHYSICAL REVIEW are respectfully submitted for approval,"—this the cover on Corson's wall? A few days later came various samples of pages set up on various qualities of paper. That selected was heavy, a far better grade (and more expensive) than that we enjoy (?) today. If we had stayed with the earlier grade, however, the space required for our holdings would have at least doubled. There were many letters to and from Macmillan regarding cuts, proofs, plates, etc. Early in the venture, the editors received a letter from the publisher; they will receive by Express Delivery a negative—"which you will find in exactly the condition in which it was received." Presumably glass and in pieces. Macmillan wrote: "It is never safe to send them through the mail, even between pasteboards." The publisher could, however, apparently still use what was wanted, in one of the pieces. No correspondence has turned up concerning a special plate for a paper by Professor Marston at Iowa. He had borrowed a couple of Nicol polarizing prisms from Nichols and wrote asking if he could keep the Nicols (we seem to have enough Nic(h)ols around here, more yet to come) a while longer in preparation of "a paper for the Journal of Physics on strain distribution" (in polarized light). A bit later he asks where to send the proof of the article and thanks for the loan of the prisms. His published paper includes the only color figure that the REVIEW printed up to the mid 1980's. Today with encyclopedias, dictionaries, and other journals coming in color, can ours be far behind? Indeed, we are catching up; as of 1989 a few colored illustrations were printed. For a price, an author can have figures colored as she/he chooses.

Book publishers were heard from. Lee and Shepard sent a copy of their Matter, Ether, and Motion by Dolbear for review, which Merritt did in the second issue. Ether *non*-drift was by now known and being speculated about, but Merritt's review makes no mention of Michelson-Morley. Ginn and Co. were pleased that Sabine's book would be in the first issue. This was a text reviewed also by Merritt. Whether this was prompted by a letter from Franklin at Iowa State (after a fellowship year at Harvard) is not clear. Franklin had urged a review of the book in a letter, for "… it may throw some light upon remarks I have occasionally made to you regarding physics at Harvard." Were it not for Franklin's regard for Sabine, he "… would like nothing better than to write a severe critical review of the book and what it stands for (by implication)." Sabine is a "splendid fellow … the best man in the Physical Department there." No indication here as to what was bothering him about Harvard physics, and Merritt does not exactly come across in the review. After a stint at Lehigh, Franklin would wind up at MIT, rather closer to the physics at Harvard than when in Iowa.

Authors also pushed their work. In a handwritten note, Bedell (name not yet on the journal cover sheet) urges haste on Nichols regarding his paper with Crehore on capacitance; it is mathematical in nature and reprints (we have no information on that aspect of the publishing venture before Volume V. It is known that Bedell made capital with his, however) would be available for the Madison A.A.A.S. meetings in August. It was a two-part paper, the first of which appeared in issue Number Two, with a footnote saying that it was presented at the August meetings. The second part of the paper was presented to the World's Congress of Electricians in Chicago, also in August, and was published in issue Number Three. The reprints were probably not available in time; it would seem that Bedell was crowding things a bit there. He and Crehore, with a note to Nichols, submitted a copy of their book for review in the first issue; MacFarlane gave it generally favorable review but not without extensive criticism.

In June, as the date for the journal's first appearance approached, Macmillan wrote Ithaca asking for a list of editors and others to whom complimentary copies should be sent; and a week later another "... have just received the first number of the PHYSICAL REVIEW and take pleasure in mailing copies ... ." They were already assembling Number Two, and had everything in hand to go to the printers "except the article by Mr. Merritt." There had been interest in increasing the run on Number One from 1000 copies to 1300 but it was already in distribution; they would put out a second edition if it seemed advisable. It did not. Merritt reported to Nichols that his paper had just gone in—3400 words with photos on the manometric flame and (presumably) his study of the sound of the vowel A, seen in issue Number Three. He never made it to Number Two. The editors were obviously very busy, and they still had their university teaching obligations. A letter of complaint to Merritt in the archives is from a student gone for the holidays to Paris. From the professor he had received all of 54% for his grade in Merritt's course; he had only passed nine hours; trouble ahead, could we do something?

It must have been an exciting day in 1893 at Cornell for the editors of the new journal and their contributors, most of whom were Cornell residents when Number One of Volume I, the July–August issue of the PHYSICAL REVIEW reached their mail boxes. It comprised a publication of 80 pages of relatively heavy paper, pages uncut. A new issue came thereafter every two months, the first year's full volume coming to 480 pages. There were a number of Departments to be perused: *Contributed Papers*, that leading off on Page One by Ernest F(ox) Nichols on "The Transmission of Certain Substances in the Infra-red," followed by one on the "Relation between the Lengths of the Yard and the Meter" by William Rogers of Colby. Others in this initial number were of similar ilk. (See the cut of the first cover taken from Corson's wall plaque used on the cover of the present memoir.) Throughout the volume, the papers seem to us today to rather support Kevles opinion—"fact finding": more infra-red (spectra of alkalies by Professor Snow), a polarization study in a voltameter (Professor Daniel), photometry with a rotating sectored disc (Ferry), freezing points of solutions (Dr. Loomis), the resistance of copper (Messrs. Kennelly and Fessenden), ditto as affected by the surrounding medium (Carhart), etc. In issue Number 3 there was Merritt's paper, not making Number 2, "On a Method of Photographing the Manometric Flame with Applications to the Study of the Vowel A," illustrated with cuts of the photographic results. Such cuts were frequent in even the first REVIEWs and appear still today; not only to line drawings have we been restricted. In fact, in Number 2, Figure 7 of Professor Marston's paper on the "Distribution of Strain in Polarized Light" is a so-so colored print, something not seen again in the journal until the 1980's. The Bedell-Crehore two-part paper, mathematical as he averred, came in Numbers 2 and 3. At the end of the year in Number 6, Editor Nichols and Franklin had the last contributed paper, save one, "On the Conditions of the Ether Surrounding a Moving Body," with references to J. J. Thompson, Maxwell, and Oliver Lodge, but no hint of the Michelson-Morley negative results, seen six years earlier. Franklin had the last paper in the Contributed Papers Section with "Three Problems in Forced Vibration."

There were other sections in the journal: one on *Minor Contributions*, one on *New Books*, and one simply titled *Notes*. At the end of Number 6, came a brief index for the year, running alphabetically from "A Geometrical Proof of the Three Ammeter Method of Measuring Power" on page 59 by Cornellians Bedell and Crehore, through *Minor Contributions* and *New Books*, and *Notes* to the "Three Problems in Forced Vibrations" by Franklin on page 442.

Minor Contributions for the year included in Number 3 "Some Rapid Changes of Potential Studied by Means of a Curve Writing Voltmeter," by George Moler (presaging faster devices some years ahead), and one in Number 5 by Louis Austin on "The Effect of Extreme Cold on Magnetism," reporting that a magnetized steel knitting needle cooled to the temperature reached by a mix of solid CO_2 and ether, rather than losing its magnetism, slightly increased it, if anything, not in accord with an earlier report by J. Trowbridge. Under the same department, Kennelly and Fessenden in "Dynamics of the Ether," reason that "... ether distortion in a moving body seems indicated by the experiments of Fizeau and Michelson ...," those of the velocity of light measured in a flowing liquid; again no mention of the Michelson–Morely experiment. Another minor contribution is of interest: an acoustic pyrometer of Fernando Sandord'; the acoustic resonance of an air column closed at one end depends on sound velocity, which depends on the temperature.

As frontispiece to Number 4, protected by a leaf of tissue, is a fine, sepia, photographic portrait of John Tyndall, who had recently died. Under *Notes* was an obituary written by Nichols, in which he harkens back to Tyndall's popular lectures, quoting the man: "... I have endeavored to bring the rudiments of a new philosophy within reach of persons of ordinary intelligence and culture ... that a person possessing any imaginative facility and power of concentration might accompany me." E.L.N., as the *Note* is signed, concludes that Tyndall possessed the talents to do this "... in the highest degree and he has utilized them for the delight and inspiration of an entire generation." Tyndall's influence in exciting the public and stimulating the rapid growth of late nineteenth century American science was not lost even on those of his own generation. In issue Number 5 there was another frontispiece portrait, and obituary by E.L.N. in the *Notes* section; this time for Heinrich Hertz. This REVIEW practice was continued infrequently for some time, the next one coming in Volume II, Number 1, for August Kundt, known for his acoustic tube strikingly demonstrating standing sound waves. In more recent times, they have come few and far between, the cumulative Index 1921–1950 (not all that recent) listing only two: that for A. A. Michelson in Volume 37 (Series II) in 1931, written by R. A. Millikan, and that for John Tate, PHYSICAL REVIEW Editor for the second quarter of this century, appearing in 1950, Volume 79, Series II. Not listed in the Index, however, are at least two others, both Cornellians. Most appropriate of all the Memorials that the REVIEW has published is that for its founder and first editor, Edward Leamington Nichols, the Memorial written by Professor Merritt, appearing as lead-off frontispiece of Volume 51 (Series II) in 1937. And then two years later in Volume 57, one for Floyd K. Richtmyer, big in American physics in the first part of the century. He was a past President of the American Physical Society, the Optical Society, and the Physics Teachers' Society, editor of some of our journals, and a great teacher, honored today in the annual Floyd K. Richtmyer Memorial Lecture at winter APS meetings. Currently, deaths of our colleagues in physics are noted and the deceased paid tribute in a column of *Physics Today*.

Beyond obituaries of noted physicists, the *Notes* section in these early issues of the REVIEW commented on such as the Helmholtz visit to America ("The arrival of Baron von Helmholtz aroused much enthusiasm among the younger physicists of the country, many of whom had been his pupils. He was accompanied by Baroness von Helmholtz and ..."), made note of new appointments ("... Frederick Bedell, Ph.D., as assistant professor of physics in the Cornell University; ... " meetings of the National Academy of Science, the A.I.E.E., the A.A.A.S., the Electrical Congress, and the like. The last named organization, quite international in scope, held meetings in Chicago in the summer of the REVIEW advent; it was to this meeting that Bedell and Crehore presented Part II of their paper. The affair could well have been a Physical Society meeting of the period ("... a gathering of electricians and physicists was brought about, the like of which has never before taken place upon the American continent."). Participants included Edison, Tesla, Anthony, Ayrton, Sylvanus Thompson, Lummer and von Helmholtz (his visit noted above; he was chosen Honorary President). Rowland and Nichols were on the program committee; Rowland indeed, was President and Nichols was Secretary of the Official Chamber of Delegates, and Webster (A.G., up at Worcester, soon to start founding the American Physical Society) was Chairman of the "pure theory" section.

And then there was in the REVIEW the *New Books* section; in the first year books by Heaviside, Ostwald, Helmholtz, Tesla, the Chemical Tables of Landolt-Bornstein, Byerly's book on Fourier Series were all reviewed. Particularly of interest to this chronicle is perhaps a lengthy review of a book by one Thomas C. Martin titled *The Inventions, Researches, and Writing of Nicola Tesla*, 8 vo., 500 pp, the PHYSICAL REVIEW commentary being written by our old Cornellian, William A. Anthony. There were many with Cornell association contributing to these early volumes.

First numbers of Volume I were pretty aware of the rank of the various authors. In the space at the top of left-hand pages would be the authorship of the article below: Professor Ernest F. Nichols, Dr. Loomis, Drs. Bedell and Crehore, Messrs. Kennelly and Fessenden, or Dr. MacFarlane and Mr. Pierce, with abbreviated titles of the paper over at the top of the right-hand pages: Study of Transmission Spectra, Freezing Points of Solutions, On Mutual Induction and Capacity, Resistance of Copper, On Dielectric Strength, respectively. By Number 6 of the Volume, the "nobility"of the authors was no longer noted and the format we have today was instituted. This may have been prompted by the suggestion of Ernest F. Nichols in a letter to the Editor enclosing "the third proof of his article for the REVIEW"; he wishes not to have the word Professor used in front of his name unless it is to be a REVIEW custom; he hopes the custom will be different. It did become so not at once but rather soon.

Some aspects of a journal are not always discovered by merely going through the bound holdings in a library. This is certainly true of the PHYSICAL REVIEW as found bound at Cornell; covers are not generally bound, nor are the pages of advertising, which were usually together at the front and back of the magazine, in first issues solely notices of books published by Macmillan. The inside of the front covers generally carried information for authors and subscribers, while the rear, inside and out, might carry interesting advertising as, for example, found in Volume II, Number 4, an ad for *Astronomy and*

Astro-physics, contents for December 1894 (inside) and (outside) the PHYSICAL REVIEW itself, Contents for Sept.–Oct. 1894 and Nov.–Dec. 1894. The inside front of the same issue gives notice of the absence of the editor-in-chief during 1894-1895. It was in going through early REVIEW correspondence that it came to be realized that advertising was a source of revenue for the journal and that original copies had best be referred to where one could. Fortunately, the Cornell Physics Department is wealthy in unbound, original, back copies, particularly those following Volume V. Prior to that only an occasional issue or so is present in the original; others of that period are in reprint form, lacking covers and advertising.

Chapter 3
THE CORNELL EDITORS

•◆•

With the PHYSICAL REVIEW safely (?) launched, one may wonder what manner of men were E. L. Nichols and his two colleagues (Bedell's name appeared on the cover page of Volume III), who steered its course over the first twenty and thirty (Bedell) years. They were different characters, with different interests and abilities. From what one reads of him, Nichols was the more introspective of the three and thought much about the place of science in society. That comes only from what is written; I never knew him, but only had spent one Sunday afternoon in his presence at Merritt's house on my first Sunday in Ithaca. It was their usual Sunday dinner to which they had invited me and my sister, already a graduate student at Cornell. His granddaughter was also there, a new physics graduate student like myself. It was a sunny afternoon and I have a picture of the venerable old gentleman, distinguished looking even at 80 years of age, neatly bewhiskered with a white beard (rather than the earlier imposing black seen in a photo with Lord Kelvin and J. Gould Schurman) sitting there in his chair, not doing much of the talking, not too interested in a first-year graduate student, a hay seed from the West who was not his granddaughter. Merritt, on the other hand, although himself near retirement, was like an elf, full of life, sparkling wit and conversation, outgoing, interested in all manner of things, enthusiastic and bubbling over. He was easy to get acquainted with and I came to know him and his likewise delightful wife, Bertha, well. Bedell I also got to know pretty well. He was rather reserved, friendly but somewhat formal, and probably the most able scientist of the three. The mix of characteristics of the men was good to have for the new journal.

Edward Leamington Nichols was born of American parents (of English descent) in Leamington, England. His father had been a music teacher but gave that up for a career in painting and was in Europe studying the medium when, during a protracted stay in England, Edward was born. Apparently the artist was not bad;

some work is said to still be extant in the minor museums. Back in this country, the young Edward prepared at Peekskill Military Academy (whither gone?) and subsequently entered Cornell. There is no indication earlier that science was in his future; at Cornell he became interested in chemistry but it was Professor Anthony who aroused him and turned him to physics. After graduation, he spent four years in Germany: at Leipzig with Wiedemann, at Berlin with Helmholtz and Kirchoff, and at Gottingen, where he got his Ph.D. degree for a thesis—"Von Gluhenden Platin Auggestrahlte Licht"— the work done at Berlin. They re-awarded him the degree fifty years later. On returning to this country, he approached Andrew D. White about a position at Cornell. But there was a depression on and no room at the time; however, White wrote Gilman at Johns Hopkins about him with the result that he went there to work with Rowland. There he repeated the latter's famous experiment on the magnetic effects of a rotating static (slight contradiction there) charge distribution, sorted out discrepancies, eliminated some errors and showed that the effects reported were largely frictional; any magnetic effect was very small. He had had some experience with Rowland's original apparatus while in Berlin. Confusion over the important experiment was finally eliminated later in the work of Cremeu and Pender. He spent some time with Edison on photometry of the new incandescent lamp. He taught then for about six years (Kentucky, two; Kansas, four) before finally making it back to Cornell as Anthony's replacement.

He inherited as permanent staff Assistant Professor George Moler, who had collaborated with Anthony in the construction of the Cornell generator. Together, Moler, Nichols, and an Instructor could handle those of the one thousand Cornell students who were inclined (or required) to take some physics. But growth set in, not only in the Department but in the physics world and in the University itself. Over the next few years it became necessary to add more staff, notably in the persons of Ernest Merritt, Frederick Bedell, Harris J. Ryan, B. W. Snow, and D. C. Jackson, and a few other Cornellians who did not become as well known. The first two—Merritt and Bedell—would become leading characters in the PHYSICAL REVIEW story. Ryan became head of Cornell's Electrical Engineering offshoot before he left for Stanford—"Cornell West"; Snow went to Wisconsin, and Jackson, elevated to head up E. E. at MIT, went to Cambridge. Moler stayed at Cornell through to his retirement, as teacher and aide to many graduate students, but not much involved further in our tale here.

After Nichols' death, in 1937, Merritt wrote the biographical memoir of him for the National Academy. For their researches in luminescence over a long period of time, the names of Nichols and Merritt became linked almost as though they were a single person. The two men became very close; following the death of his wife, Nichols took Sunday dinners at the Merritt's; to the Merritt kids he was known as "Uncle Ned." Merritt has this to say in his memoir about the man:

> *Nichols was a pioneer in several branches of physics. Much of his work called for manipulative skill of high order; all of it called for ingenuity in meeting new problems. But when an investigation reached a point where high precision was called for he was ready to go to something beyond ...*

> *One of his most important contributions to American physics was the indirect influence he exerted through the students who received their inspiration from him and who later entered the field of college teaching or industrial physics. At the time of retirement (1919), the heads of departments in thirty-five colleges, fifteen of them state universities, were men who had received their physics under him.*

No wonder the REVIEW was able to find contributors.

He, with sundry coauthors, wrote a number of books, now largely to be found only in library outlying annexes: *The Galvanometer* (certainly covering the principles of Anthony's big instrument), a *Laboratory Manual of Physics and Applied Electricity*, the three volume *Elements of Physics* with Franklin, *The Outlines of Physics*, and numerous long tracts on luminescence. He was president of the A.A.A.S. in 1907, of the American Physical Society (1907-1909), and of the National Honor Society of Sigma Xi (1908). He received in 1928 the Rumford Medal from the American Academy of Arts and Sciences of Boston, in 1927 the Cresson Medal from the Franklin Institute, and in 1929 the first Ives Medal of the American Optical Society. Planck also got a medal with him that day at the Franklin Institute and spoke on "The Reality of Photons"; he had come around by then.

Nichols was quite a forward looking man for his time. For those of his Department interested, he held weekly evening seminars at his home. His views would please revolutionaries and environmentalists alike today. In his retiring address as president of A.A.A.S. on ``Science and Practical Problems of the Future," he sees three or four problems in the solution of which science will be needed. He deplores the waste he sees in the destruction of the bison, forest, fish, etc.; the prevention of such and ``other manifestations of corporate greed" will be solved through laws, but science will be called upon for guidance. He sees the eventual exhaustion of our coal and petroleum supplies. He wants to see the utilization of solar energy in ways other than by natural storage. But he wants the seeking of fundamental knowledge to go ahead and practical results come; the work of Joseph Henry is cited. He complains that when a man becomes productive, he is shoved off into a presidency or directorship. He hopes for discipline and reason in the front office, quoting the "testy remark once made by an eminent scholar: 'You can't run a university as you would a saw mill.' " And: "We need not merely research in universities but universities for research." Modern notions.

In a *Scientific Monthly* summary of a speech he gave on the occasion of Ezra Cornell's 114th birth anniversary, he deplores tradition, tradition as such. He was not unhappy over the student activism common in Europe but "Nobody wants a university full of anarchists. They are a nuisance—certainly a nuisance ... ." Why aren't our students active? Tradition. Let's have some reason.

It is perhaps worth quoting at greater length some excerpts of the response he made at his retirement dinner (Tickets: $1.10; steep enough for the participants, in 1919. Entree: Roast beef and brown gravy. In a reversal of today's ordering, roast chicken could have been provided at $1.50 a plate. The arrangements committee were clearly on the side of frugality.) It was also the occasion of the semicentennial celebration of the University. He tells of his "dream." Cornell had had half a century showing the way in a daring experiment in education and "... we glory in that fact. Yet she, who has boasted of her freedom, is fettered like her sister institutions throughout the land by one bad tradition; the tradition of school masterdom. My dream is of a Cornell that shall be first to break away into glorious freedom that surrounds us into the glad Bohemia at our very doors. We get whiffs, we are not dead, we struggle feebly." He saw the universities as the "workshops" for the urgent tasks ahead—great effort was called for, there cannot be dilletantism. We need freedom, devotion, and the scientific spirit— they define the *university*. It must declare the advancement of knowledge to be the prime, its supreme function, and make it go. And then he winds up: He hopes to be around and still curious, having *"the privilege of watching the wheels go around, of that is all I feel I can do or ever have done. It has been delightful—unspeakably delightful—that life which comes from the study of science. What I would like to say and cannot, is that you must not be content with the things the generation that is passing away had to be content with. It is for you to do greater things and more important things than we have ever done. The things are crying to be done, and the world is crying out to have them done. If Cornell is to be what we all hope and believe she is to be, it can only be through endless strivings of the imagination, through ceaseless labors and great creative art. It can only be by the highest efforts of everybody who has a mind to do anything whatsoever. Then we can look back upon the crude efforts of those who went before and, while we may smile, we may at least believe that they looked forward to the things they could not accomplish but which you shall accomplish."*

To some ears, and not only those of Cornellians, this is telling testament, and we should keep it in mind as we peruse paper titles in the early PHYSICAL REVIEWs.

Withal such shining idealism, in which Nichols was not alone. (Physicists of the time had not gone into the field for lucrative promise.), it is surprising that so few worried about really fundamental questions, or at least made any progress on them. There is in the Cornell Archives (Nichols' file) a copy of a talk Nichols gave in Albany at some Institute over there shortly after Roentgen's discovery of X-rays: "The Photography of Concealed Objects: the Roentgen effect." In the Archives are reproductions of X-ray pictures of keys, gears, etc., taken by Professor Moler at Cornell in the days following word received of the discovery. In the Merritt collection at the library there is the negative and print of the X-ray Moler made of Mrs. Ryan's hand atop a 25¢ coin; exposure 15 minutes, three feet from a small hole in front of the tube anode. Lots of places had the means for doing this sort of thing; gas discharge tubes and high voltage induction coils were common. But it would be two years before the REVIEW would carry a paper on X-rays, with R. W. Wood's looking for wave characteristics and reporting a "high intensity" source.

Ernest Merritt, as can have been surmised by now, was a student of both Anthony and Nichols, graduating a year before Anthony's departure from Cornell. He earned a Master's degree and became an

Instructor on Nichols' faculty and an Assistant Professor then in 1892, the year in which the notion of the REVIEW was broached. He was thus active in the Cornell department for nearly half a century and a vigorous force in physics for only a decade less, as we will see. For some oldsters, the days of Merritt seem not so long ago; he died a few years after the end of World War II, still a charming and concerned individual. He and his wife had helped his friend Max Planck through a distressingly hard time after Germany fell.

Unlike Nichols, Merritt had leanings toward science at an early age. The interest was astronomy. The *Indianapolis Journal* in 1879, in an article entitled "The Boy Astronomer," reported on the fourteen year old Ernest "whose eager study of the stars may some day make him famous"; how during an illness he was inspired "to translate the mighty secrets of the universe." He had purchased from a local star gazer a small telescope, housed it in his modest observatory, equipped with "transit, desk, and stove (!) and a *stereo plasa*, an invention of his own," not patented presumably, nor preserved, and thus lost forever. Earlier, in seeming forecast of things to come, he was into journalism, non-scientific, reporting, it is hoped, in less florid style than that of the *Journal* reporter, who further tells us that at the tender age of eight, Ernest had edited, scripted, and illustrated "The Sea Breeze," his own paper about the size of a man's hand. Later with a small hand press he had acquired, the sheet became larger, "printed, stitched, and bound" under a new name, "The Mountain Echo"(!). All this in Indiana, where he grew up.

It is not clear when journalistic interest became astronomical interest. His interest in physics was apparently kindled when he came into contact (like Nichols) with Professor Anthony and his popular lectures. In spite of flunking the first course in the subject, he fell for it. The failure is odd in view of his earlier interest and his having received the mathematics First Prize from the Indianapolis Classical School. In spite of the setback, he took more physics. While he got his undergraduate degree in Mechanical Engineering, he took a Master's degree in physics. In 1889, having seen him as a student, Nichols appointed him as an instructor. He subsequently rose in rank and would become Nichols' successor as Department Head, to retire from active duty in 1935. It was a privilege to be at his retirement banquet and hear him reminisce; he told of his having looked forward to the day when he could be Professor E. Merritt, Professor Emeritus.

Merritt was something of a motion picture enthusiast—that is, he made his own. Cornell physics must have had a yen for this sort of thing. After all, Ithaca was in the early days a world movie capitol; Pauline was put at her peril in various places locally. Professor Moler anticipated Disney: he took moving pictures of a skeleton, resurrected from the Anatomy Department, performing a wild dance. He had rigged it up and manipulated it with strings, arms and legs flinging about and head bouncing off and on. He took daily pictures on movie film from a fixed site, of the construction of a new building going up on campus. Run through a projector they showed the building going up at a furious rate. Run backwards, time reversal. Merritt told of going out in an afternoon of snow or freezing rain with his camera, heading for the foot of the long State Street hill in Ithaca to make movies of hapless and frustrated motorists sliding down the hill with degrees of freedom to which they were not accustomed, and of other optimists slithering around, spinning their wheels in attempt to gain elevation. The resulting footage, shown at a physics party, was pretty hilarious. That was in the days before snow tires and before streets were salted down at the slightest provocation. Cars lasted longer in those days too.

There was another film he took great pleasure in showing. It was taken from a boat following the Cornell crew rowing in some regatta. In the foreground was hanging a rope fixed to the vessel on which he was traveling, the shell off in the background, somewhat aligned with the rope. It was interesting and instructive to the uninitiated to see the back and forth motion of the racing shell relative to the fiducial mark on his constant velocity platform, with each reach back of the oars for the next stroke, the shell surging sharply forward. After a visit by Max Born to Ithaca, Merritt wrote him apologizing for the bad weather, sending him pictures of a more benign Ithaca and enclosing a strip of film "of Millikan, of a new kind being developed in a laboratory not far away from here (undoubtedly Eastman Kodak); the voice is reproduced as well as the picture." Millikan has been lecturing at Cornell and talking movies had been made.

Long after he became a member of the faculty, for some gathering or publication, Merritt prepared a short dissertation on liquid air. That was the physicists' refrigerant (after dry ice and ether) at the turn of the century, and Cornell had come into possession of a monstrous liquifier, a gift to the Physics Department. The old machine came into yeoman service of the city of Ithaca. There was a typhoid epidemic raging, probably that which Morris Bishop notes as arising from unsanitary conditions along one of the creeks flowing through town. At any rate, in the particular epidemic, the hospital ran out of oxygen, which was used in treatment of patients with the disease. It was a weekend and no trains ran between Syracuse and Ithaca on Sundays (unbelievable today, one used to be able to make that trip) and the oxygen supply was in Syracuse. The University was asked for help. The chemists set about trying to make the gas by breaking down oxides, a pretty slow process. Merritt tells of Nichols' suggestion that liquid air be cranked out with Physics' machine (cranking out is approximately the correct term for it if one knows the early machines) and then allowing the nitrogen to boil away, leaving the oxygen, which could then be got into cylinders. And so it was. The liquifier ran full tilt, the product put in "pails" and allowed to boil largely away, the remainder being sent in cylinders to the hospital, tiding them over until the following Tuesday.

In the same dissertation, Merritt is obviously enjoying relating the following tale; it typifies his enjoyment of life. It seems there was to be a popular lecture on liquid air given by someone shortly after the appearance of the commodity on the Cornell campus. The lecture was indeed popular; so crowded was it that Merritt could not gain entrance to the large lecture hall and so he went around to the back into the apparatus room behind. There he could barely see the lecturer, but enough of what was to transpire. There was a fire extinguisher hanging from the wall and immediately below it, not yet disposed of, was a pail of old light bulbs (carbon filaments, no doubt) which a custodian had but recently taken from the overhead banks of room lights, replacing them with new bulbs. All was now bright and cheery. One clearly interested spectator, sitting in an available space on a table near the wall, squirmed about somehow, the better to see the demonstration, and managed instead to knock the fire extinguisher off the wall, dropping it into the pail of bulbs, upsetting it and spilling the contents. The fire extinguisher went into action, spraying a mix of H_2O and carbonic acid about in all directions. It was clear to the audience that liquid air was loose and about to freeze everyone to stone. The spectators scrambled away, trampling the old lamp bulbs lying about on the floor. The resulting explosions, indicated to the crowd that the liquid air was indeed going off and more panic resulted. Positive feedback. In the pandemonium, no one was injured, fortunately. Merritt notes that "I never succeeded in seeing or hearing the lecture except this, but it was sufficiently interesting."

Merritt was a good teacher. I well recall his final lecture before retirement, his good wife watching from the back row, looking down on the podium. It was largely a demonstration lecture course of phenomena in gaseous discharges. It was fun to watch him at work.

He'd set the experiment to going; the mechanical pump would take off glubbing away, the vacuum develop, then the first gaseous glow followed by whatever it was that was supposed to happen. He'd look up at his audience and beam, seemingly as astonished as the audience at what was taking place. So it was in his theory courses it was said, always surprised at what had resulted from the analysis. An engaging way; I can still see him down there.

Beyond his normal teaching obligations, the first fifteen years as physics professor were spent in various research activities resulting in publications on the likes of glow lamps, flames, photoelectricity, batteries, radio waves and oscillations, cathode rays and (of course) the manometric flame; the second fifteen were spent largely with Nichols on luminescence; and the last fifteen in the Department chair. And during most of that time, active in the American Physical Society. Merritt, as Secretary of the Society, carried the load of its operations during the early critical years, then as President and council, serving continuously as an officer of the Society for just short of fifty years.

His interests are seen to be widespread. In a remnant of his early astronomical bent, he combined that with radio. He got involved with a 1925 total solar eclipse expedition, wondering about a change in radio propagation at totality. Others had undoubtedly done similar measurements; like cosmic ray investigations in its early days, there was usually a chance for a trip to exotic places. Unfortunately, this one of Merritt's took him only out of his house, for the path of totality was right across the Cornell campus. There is a beautiful picture (a copy on Corson's office wall) of old Boardman Hall and library tower, at the south end of the spacious Arts quadrangle, of sequential exposures taken during the eclipse, snow on the ground, with the totally eclipsed sun and its corona at the middle of the sequence, directly over the Boardman building; it must have been near noon; ideal. One presumes Moler to have been the photographer. The event, which lasted all of two minutes officially, for weeks actually, according to Merritt. "What a show," he noted. Whatever was learned has not been preserved.

He had a continuing interest in radio wave propagation, dating from World War I days, if not earlier. At the end of that conflict he suggested using radio to locate ships. (But the first suggestion for such has been assigned to many experimenters. Breit and Tuve at the Carnegie Institution pioneered *pulsed* radio ranging with their measurements of ionospheric height. Based on that technique, various places here and abroad were developing radio detection and ranging during the 1930's. But old Nicola Tesla antedated them all; he suggested back in 1900 that radio could be used for detection purposes.) Merritt saw the connection between aurorae, magnetic storms, and propagation effects, as did others. In a shack atop a wood tower erected on the Cornell campus, he received and recorded radio signals, reporting on the measurements at a National Academy meeting in Schenectady, as interaction between ground and sky waves. (More notably at the same meeting, Gurney discussed nuclear spontaneous disintegration and barrier penetration, "a natural consequence of the new wave theory of matter." This was first understood by Gurney and Condon, and independently by Gamow, who is generally given the credit.)

By this time, the new physics was upon us, American physicists were still going to Europe where it was breaking fast. Cornell physicists over there wrote back that it was tough; one had to learn something new every week. In a memo to his Dean, looking for more support, Merritt discusses the problem of why European universities were in the forefront of the new developments. Several reasons he cites: They give small time to undergraduate courses, they have excellent material equipment and support, they have small teaching loads, and no administrative chores. It was undoubtedly not quite that simple but there was certainly some truth to his argument. In an earlier report, he worries about why we don't get more physics students and then points out that possibly students go to engineering, only to discover physics too late. We should stress physics as educational more than informational, get across the spirit of science.

In 1932 Merritt was in Europe on a sabbatical leave, checking out the territory, visiting around—Leiden, Gottingen, Berlin—his eye out for new staff. And he made notes which are of some interest. In spite of Kammerlingh Onnes' death, the low temperature work in Leiden is going along fine; under Keeson and deHaas, he was shown solid helium at $0.8°K$ (at 60 atmospheres), first obtained a few days previously. He spent a day in Haarlem with Prof. and Mrs. Lorentz—"very pleasantly indeed." From Gottingen: he thinks Born's decision (not to come) is final (he had been offered a position at Cornell), even if Franck were called to Ithaca; and Franck might be even more valuable—but "it would be grand if we could get them both." (Franck did spend a summer and give a course in the Cornell department before taking up his first semi-permanent position, at Johns Hopkins, following his exodus from Germany. He was another gentleman in the real sense.) Merritt saw an "excellent lecture corresponding to (Cornell's) Physics 3, by Pohl at 7 AM." (!) Was delighted. "Pohl is doing nice work on crystal conduction produced by light." (Before or after the great experiment on the Hall effect in diamond done with Gudden in which the unclamped poles of an electromagnet slammed together crushing the huge borrowed stone?) He is favorably impressed with the younger men such as Oldenberg and Rupp. "Did not meet Jordon. Pauli is impossible." In Berlin he had dinner at the Planck's; von Laue was there, but Merritt was unimpressed—"he talks too fast—even the Germans can't understand him." Berlin and Gottingen are much interested in Schrödinger's recent articles on quantum theory; Merritt thinks that Planck regards it as the most promising step yet. It was an interesting discussion at Planck's between Planck, von Laue, and Franck; the last sees Heisenberg as largely formal, and Schrödinger's approach more capable of physical interpretation, but the real solution will be something different still, not yet reached. We may add, interpretation perhaps still not yet reached.

The Merritt's were of Quaker persuasion but with the entry of the U.S. into World War I, Merritt lost little time in getting involved, urging that physics be used in combatting the submarine, and in the location of concealed artillery. Even before our entry, he had written the Navy Secretary proposing means for signaling to and from submerged vessels, and urging the establishment of a Naval research laboratory. Generally speaking, however, it was a chemists' war and physics played no great role. There was a committee, headed by inventor Thomas Edison, which undertook the task of identifying the area where physics and technology could make a contribution. Edison had little regard for fundamental work, as we've noted, and not much can be said of the committee's benefits. Physicists were pretty unhappy over the committee's disdain and for being essentially disregarded. A letter from Merritt protests the absence from the committee of any physicists from academia; Millikan concurs in the protest. Astronomer George Ellery Hale seems to have been the big spokesman for science involvement at the time. Kevles relates the War picture interestingly in his book.

At any rate, Merritt went to the New London station, where submarine detection was being pursued. The Cornell Archives holds considerable Merritt correspondence of the period. There are letters to and from Millikan, Michelson, Jewett (head of Bell Labs, to come) and others; nothing seen to or from Hale surprisingly. The name V. Bush appears signed to some correspondence; his would be the big name twenty–five years later. There are reports of tests carried out in the New London waters; these involved hydrophonic detection and magnetic induction techniques, and they were not unsuccessful. Some two decades later, after Pearl Harbor, Merritt wrote Frank Knox, Navy Secretary, asking why the harbor had not been protected by an induction cable on the floor of the harbor entrance; it was not a totally useless tool. Merritt is later pleased to learn that a cable was in place, it did detect submerged vessel penetration, but in the chaos of the day there were too many other things going on; at that, one or two enemy submarines were sunk. Beyond this, he was quick to write Bush asking him to let physicists know what needs to be worked on so they can get at it, and he writes Lee DuBridge at MIT of various techniques and devices possible.

While they were into retirement by 1940, the Merritts were not exactly disinterested bystanders to the conflict then raging. They were very active in the Bundles for Britain and, after the war, with the American Friends Service Committee in relation to relief for friends in Germany. In particular, with Professor Cope of Pennsylvania, they made an effort to get in touch with Planck and his family to learn of their welfare and to get CARE packages to them. Merritt was fairly close to Planck. In the Cornell file are the neat notes that Merritt took of Planck's courses "Warmtheorie" and "System der Physik." There is a New Year's greeting from Planck, 1939 the year, when things must have been gloomy for the great man. He stood up to the Nazis; one of his sons was later to be shot for involvement in the unsuccessful plot against Hitler. After the German surrender, the Merritts were able to get in touch with the Plancks. There are several letters from Planck's wife, Marga, and son, Herman, expressing their appreciation for what had been done for them. Planck himself does not write in this period; he is very feeble and failing. He had taken to his bed and considered his life over.

There is in the Merritt file a nice letter concerning Planck from Yale's Walter Miles, which has little to do with the PHYSICAL REVIEW; Planck actually never published in the journal. But what the letter relates cannot fail to be of interest to physicists. In connection with a Newton Celebration of the Royal Society in 1947 there were many distinguished guests. At the celebration, in London, a page in a loud voice announced the honored guests, one by one in turn, from this or that country, names appearing on the official list. The list was finished and the page then went on and announced "Professor Max Planck—from no country." There was a standing ovation as Planck slowly rose and was helped to greet the president. The Society had made special arrangements to get him there, going so far as to charter a small private airplane. Miles regrets Merritt's not being there to greet his old friend and professor. The "from no country" was the page's error but Planck had not been listed. This was rectified the next day with the citation "from the world of science." Miles goes on to tell of a small dinner he gave for the Plancks to which he invited Bohr. Planck had hoped for a small dinner conversation with Bohr. Miles writes that things went pretty slowly at first but that they warmed up and overall it went very well.

Thus Ernest Merritt. He died in 1948.

Frederick Bedell was quite a different character. He was reserved, of fair stature, stern of face, more limited in his interests but, from this perspective, rather the better physicist and analyst than either of his two colleagues; he went into his interests in a deeper way than either Merritt or Nichols; his researches were more far-reaching than those of the other two, some of his ideas and concepts permeating our lives today. He was admirable leavening as third REVIEW editor. Raised in Brooklyn, he got his undergraduate degree at Yale. While there he read some published work done at Cornell as a Senior thesis and decided that if that was the kind of thing they did up there, Cornell was the place for him in graduate study. He tells how there were no set course requirements, there was no graduate faculty, no dean, "... I simply did as I pleased, taking my oral examination (with Thesis) in 1892, the first year Cornell gave that degree in Physics." And he stayed on, in the next year joining Nichols and Merritt in editing the PHYSICAL REVIEW, and continuing for a decade after they had left it in his hands under the American Physical Society.

His contributions to the use and understanding of electric power were the equal of any electrical engineer of the time, even including Edison, with whom he experimented on a three-wire system from a single transformer. He was one of the very early expounders of AC circuit theory and in large measure was responsible for the fact that today, almost universally, power is distributed as AC rather than as DC, which Edison favored. Bedell understood the transformer and three phase circuits, and he could measure AC power in the days before wattmeters (see that paper in the first REVIEW). He introduced the use of vectors and the circle diagram in analysis of the transformer and synchronous motor, writing a paper with Harris Ryan on the action of the single phase synchronous motor, measuring rotor phase difference between generator and motor by simple optical means. (How many of us today can explain the motor; Professor Weisskopf some years back looked in vain in his faculty for some one who could!) He presented with Albert Crehore (whose sister, Mary, Bedell married) to the Institute of Electrical Engineers the first significant paper on AC theory. Together they wrote the first real textbook in Electrical Engineering, *Alternating Currents*, which was translated into many languages. It was for years the standard reference and text on the subject. *Direct Current and Alternating Current Manual* followed. He wrote *The Principles of the Transformer* and two others, *The Airplane* and *The Airplane Propeller*, outgrowths of his involvement in World War I in his course on airplanes. In 1894 he wrote with G.E.'s Steinmetz the first paper on reactance. Again with Crehore, in 1892, he called attention to the limitations of the telephone in long distance communication imposed by the difference in the "decay" of high and low frequencies, which they pointed out could be reduced through the introduction of "self induction" in the line. The theoretical deduction was substantiated later by Pupin at Columbia; he did just that and is generally given credit for making long distance telephony possible with his "Pupin loading coil." Bedell understood the importance of wave form in AC and was led eventually to the use of the electronic oscilloscope, first demonstrated by Braun. Bedell was the first to use the multiple trace oscilloscope. The linear time base (from the voltage rise on a charging condenser, stopped by the breakdown of a neon glow tube across it for the flyback and restart) for the oscilloscope was known, but Bedell and his student H. J. Reich were first to stabilize the sweep, which was to make the instrument (they were the first to use the term "oscilloscope") useful in providing stationary wave forms on the screen. This was patented and the Burt Scientific Co. in Pasadena—(Burt was his son-in-law)—manufactured an oscilloscope incorporating this feature. There were a few of these still around the Cornell Physics Department in the mid-thirties. The comparison with modern scopes is striking. The oscilloscope was in a wood cabinet with folding front doors, and included, besides the sweep circuit glow tube, three vacuum tubes, two of which were in use at any one time. Through a peep hole in front, the operator adjusted tube filament temperature visually. The specifications on the vacuum tubes (UX-201A's, a version of the UX-201, the first commercial tube for home radio owners) cautions: ``Filaments should always be operated at the lowest voltage which will give satisfactory results; **Avoid Putting the Plate Voltage on the Filaments**'' (!). Beam focusing in early cathode ray tubes was achieved with ions created by the beam acting on the residual gas. In consequence, only relatively low frequency signals could be displayed. In the mid-thirties, DuMont in New Jersey sent Bedell some new, highly evacuated cathode ray tubes in which electrostatic lenses played a major role; ion focusing was almost nil. By this time radio tubes had gone well beyond the UX-201. Bedell and his student Tom Goldsmith created fast sweep circuits for the new cathode ray tube, making it possible to look at high frequency signals. No more 201's, no more neon glow tubes. Goldsmith went from his Ph.D. to development at DuMont, whose commercial scopes became almost the standard up to the time of World War II.

Being hard of hearing, Bedell invented a bone conduction hearing aid, and the "Deaf Speaker,"a reproducer for the hard of hearing, using the same principle; and he was among the first to apply impedance matching to such devices. Merritt, hard of hearing himself, was a recipient of one of Bedell's "Deaf Speakers," a little box connected to the radio, out of which came a stiff cable with a clip holding at the far end a disposable tongue depressor. Held in the teeth, one could hear the radio. Merritt got quite a lot of pleasure from this unit.

At the 50th anniversary of that first "pioneering and classic" paper presented to the A.I.E.E., his work was cited at the Institute meeting in Victoria, B.C. Bedell's contributions were enumerated and he reminisced a bit: "In 1890, alternating current was just plain freak; it did not follow Ohm's law and 'clogged' itself in the circuits. Everyone was afraid of it." The first installation had been a 4000 foot line carrying 500 volts, which he contrasted with the 275 kV lines in construction for Boulder Dam. Voltage had been stepped down to 100 volts with a transformer, itself then new. Seen in the distribution of electricity through a community at 500 volts or more "was grave fire and life danger." A bill was introduced in the Virginia legislature limiting AC voltage on lines to 200 volts because AC was regarded as "more deadly than DC." He told how the early workers were worried about wave form, "whatever that was." Harmonics in AC could be dangerous; inductive effects were worse so that high voltages could occur, and often did. The paper with Crehore answered the question of why current was "clogged" in some circuits while voltages jumped to distressingly high values in others; it treated both the transient and the steady state; in the paper, the symbol j was introduced in AC circuit analysis for the first time.

He enjoyed recollecting the old days. He tells how at the 1876 Exposition, where he saw Cornell's generator generating, and Bell's telephone speaking, he never imagined that he would one day be

associated with Bell, Edison, and Anthony, later to be at a laboratory with "Edison verifying the possibility of having a three--wire
system from windings on a single transformer" or "listening on
headphones with Bell in New York as he first telephoned Ryan in
California," and with Anthony "using equipment he had built."

He was always pretty formal, as indicated by correspondence
during periods in which the real Chairman of the Department was on
leave and he acted as substitute: Dear Mr. Gibbs (a later Chairman),
Dear Mr. Merritt, etc., Sincerely, F. Bedell. There would be some
pleasantries and then the business at hand: he has "recommended
the appointment of L. P. Smith in Theoretical Physics; he has made
some assistant appointments, etc." He had an eye for the dollar and
he was pretty sharp. According to one Cornell Professor Emeritus,
who was a student in Bedell's course on AC circuits, Bedell would
come to class loaded with bundles of reprints as required reading
(why not?), selling them off at a price thought to be a bit steep. He
was frequently called in as consultant, and he had a number of
patents to his credit, and he had income from his books; so he was
not the poorest member of the Cornell group. He recalled how at one
point in his early consulting career he was called in as an expert witness in a patent case, not one of his own. This was on a step down
transformer for which the patent had been disallowed; any fool
knew you couldn't get more current out of something than you put
into it. One assumes he handily earned his witness fee.

When he stepped down as editor of the REVIEW in 1923, he and
his wife took an automobile trip in a circuit around the United
States. Perhaps his stepping down was occasioned by the trip; or
vice versa. It was a bold adventure, "highways" being what they
were (or weren't) in those days and the machinery and motel (tent)
accommodations being what they were back then. She wrote it up in
a modest book, *Modern Gypsies*.

He died in 1957, but I find no notice or necrology for him (or for
Merritt either for that matter) in the PHYSICAL REVIEW. Strange.

Finally, we have on the PHYSICAL REVIEW staff in the early days,
two women, Helen Lyons and Aloysia King, Nellie and Al to
Cornellians, two great ladies, very important in management of the
journal and the Cornell Department. Nellie served for 57 years,
arrived as "a mere slip of a girl" as she would put it much later. Al
arrived on the scene some few years after Nellie. Bedell, in his recall
of the old days, praises "the unsung heroes, Nellie Lyons and
Aloysia King, to whom we owe so much. Can you imagine (he is
writing to Professor Harley Howe) how the Department and the

PHYSICAL REVIEW could have been run without them, with only an occasional secretary for a few hours to help out and no organized office? Looking back to a period in the early years, Merritt I believe in Germany and Nichols beyond his limit and for a while away, dumping the REVIEW in my lap in my rooming house on Buffalo Street. It is hard to see how it could be done even with me working nights and holidays and when on vacation with the REVIEW in my trunk. But survive it we did. Then what a blessing was Rockefeller Hall with Alice (he slips) and Nellie." Nellie herself could recall much of the old days, and considerable of the Department gossip of later times. She would tell of their being but one telephone in the building and having to chase all over the place hunting down this or that professor wanted on the line.

And there we have the principle players in the "Cornell days" of the PHYSICAL REVIEW. Important also, of course, were the many colleagues and past associates of the editors who must have been badgered for articles to fatten things up between the journal covers. Kevles in speaking of the REVIEWs "seeming Baconian eagerness to publish the report of any experiment" (lacking prestige, losing much good work of younger men to European journals, or in the case of spectroscopic articles to the *Astrophysical Journal*, coming up in this "history") says that about 35% of the articles were by Cornell Ph.D.'s and faculty, who accounted for 21% of the authors—this in the first decade or so of our journal. For its financial support, for the industry of the REVIEW proprietors, and the interest and support of people trained there, Cornell University served American physics— indeed world physics—well and deserves much gratitude from our profession.

At one point in time there was official indication that another man had been in on the founding of the PHYSICAL REVIEW, and in a way he was. It was in 1934 or thereabouts that Merritt wrote then Editor-in-Chief, John Tate of Minnesota, over his unhappiness that nowhere in the REVIEW was credit given to its founder. Should we not carry the information that it was founded by Edward L. Nichols? Such practice was carried out by the American Mathematical Society in its journal—why not the American Physical Society with its PHYSICAL REVIEW? The suggestion was taken and on the cover sheet for each volume subsequently there appeared in gothic letters under the journal name the note, "Founded at Cornell University in 1893 by Edward L. Nichols." A little later the information appeared on the inside front cover and the lead page of each issue. Fine. All went well until 1959, under the editorship of Sam

Goudsmit, when the inside cover accreditation was given to E. N. Nichols. Granted that there were a number of Nicholses mixed up at Cornell with the REVIEW, there does not seem to have been one E. N. It was assumed that someone intended the founder to be E. F. Nichols, for there was such a noted physicist. Indeed, we have seen that the first article in the first issue was authored by Ernest Fox Nichols, best known for his work at Dartmouth with Hull in demonstrating and measuring the pressure of light, done independently of Lebedev in Russia, and probably the outstanding real scientist of the early Cornell days. Our E. L. tells of his introduction to E. F. During his time in Kansas, he, Edward, went out to Manhattan (Kansas, that is) to give a lecture in the Kansas State College chapel there on science, physics, or some such. A few years later, as Professor of Physics at Cornell, he was surprised one day by the arrival of two young Kansans hoping to do graduate study with him at Cornell. One of them was Ernest F., who had heard that lecture out there and decided that Edward L. was the man under whom to pursue graduate study; which he did. What became of his friend is unrecorded. Admission policies to the University must have been somewhat more cavalier than is the case these days. With E. L.'s interests in thermal radiation, it is not surprising that E. F. went into that area, as his first paper attests. He remained in the field for much of his productive career; witness the pressure of light, the first detection of heat from a star (Arcturus), the beautiful demonstration of the electro-magnetic character of his radiation in the selective reflectivity of a crossed grid ruled in a metal film, his little square resonators doing just what they were supposed to do, a demonstration that E. L. ranked right up there with what Hertz had done. He earned his degree at Cornell after an intervening two years in Berlin with Reubens. After another year in Ithaca as "Fellow in Physics," he went to Colgate and then to Dartmouth. Then it was to Columbia, and then back to Dartmouth, this time as its President. He left that for Yale and that in turn for industry at Nela Park. MIT called him to its presidency but ill health cut short his tenure. After recuperation it was for a return to Nela Park and a less strenuous life. He surely moved around. He died addressing an audience at the National Academy.

So there was another Nichols for the cover accreditation, however wrong his middle initial. The writer noticed the error one day and dispatched a letter of dismay to Goudsmit, who promptly replied apologizing for the mistake and appreciating someone's reading the journal from cover to cover. The true founder was thereafter restored.

But the error wasn't so far off the mark; at least E. F. had the first article, which made it almost okay. Somehow, however, acknowledgement that the journal was founded by anyone at all, anywhere, at any time, has disappeared. More's the pity. Perhaps with the centenary of the journal we might have it restored. Anyway, the true founder lived to see the PHYSICAL REVIEW well established with a world wide reputation and, in his last years, he was given recognition in each issue for what he had done.

Chapter 4
FIRST YEARS

•◆•

Following the initial issue of the PHYSICAL REVIEW, others came in fair regularity every other month, a new volume beginning with the next and each succeeding July issue until, with Volume V, in 1897, the journal became a monthly, almost, with new volumes subsequently beginning with the January and the July issues, at least five issues per volume. There were glitches: manuscripts not arriving on time, lost cuts, an occasional subscriber missing his latest copy, and so on. But by and large Macmillan and the "conductors"of the journal managed to get things going, with obvious support from the "contingent" of Cornell physicists, Cornellians still there and Cornellians gone elsewhere, support much manifest through the first series of the REVIEW. There was undoubtedly a fair degree of persuasion applied in the Cornell Department to get contributions. Many of the papers published seem rather quaint to us today. "We may smile"at the crude efforts of our predecessors, to paraphrase Nichols' valedictory, but it is not in a spirit of disdain, condescension, or smugness that herein we will cite some of these, as well as others of genuine interest or of real historical importance.

By the time of Volume II, a new editor was added to the first two, but it was not until Volume III that the cover sheet indicated that the "*Journal of Experimental and Theoretical Physics* was conducted now by Edward L. Nichols, Ernest Merritt and Frederick Bedell." The latter had been awarded his degree and had become a Cornell physics assistant professor, colleague of the other two. The notice of triumvirate editorship appeared thereafter in each issue until 1913, when the American Physical Society took it over under the sole editorship of Professor Bedell and an Editorial Review Board. In 1894 Bedell apparently found his two new roles pretty exciting, neglecting to turn his thesis into headquarters. A letter from the Cornell Registrar, David Hoy, a year or so later asks if "in the excitement he has forgotten" to do so. A penciled notation at the bottom of Hoy's inquiry reads: "Deposited 25 copies, Oct. 29. F. B." Things back then seem to have been pretty relaxed in the graduate school of the upstate institution.

With the added editorial staff it appears that Merritt could be (and was) on a leave in Berlin during the academic year, 1893–1894, the better part of the journal's first year, listening to Planck's lectures, passing, as he has written, the Kaiser frequently on Unter den Linden Strasse, and generally becoming awakened and broadened. He was not unmindful of his journalistic responsibilities at home, however, writing for example to Nichols of Kundt's death; the frontispiece photo and a memoir by E.L.N. appeared in due course, Volume II, Number 1 (Issue VII of the succession—we aren't going to keep track of these!) of the REVIEW. The following academic year, 1894–1895, found Nichols' himself on sabbatical leave, sending letters, cards, and at least one cryptic wire to Bedell: "Approved." On the inside of the front cover (Volume II, Number 4, previously also referred to) is the *Note*: "During the absence of the editor-in-chief in Europe (1894–1895) the REVIEW will be conducted by Ernest Merritt and Frederick Bedell. Correspondence relating to contributions should be addressed to the editor, at Ithaca, N.Y.; subscriptions should be sent to the publisher." There follows the matter of cut pages and abstracts mentioned earlier. In the same volume later another memoir by E.L.N. was written on von Helmholtz, who had also died. With his two memoirs, the journal included a two-part translation (from German) Nichols made "On Weber's Theory of the Glow Lamp" (Edison's incandescent creation), being a paper on the spectral distribution from the lamp; there was a paper with Mary Crehore on "Studies of the Lime Light" (light sources for early projectors and flood lighting was obtained by directing a flame against a block of calcium oxide; presumably that fine projector of Andrew D. White's acquisition was a true electric arc), and another with a second Mary (Spencer) on temperature effects seen on the transparency of solutions. There was a *Minor Contribution* as well: "A New Form of Spectrophotometer." In addition a third Mary (Chilton Noyes), who wrote on Young's modulus of heated piano wire, worked under and was assisted by Nichols. These three Mary's were the first of their sex to publish in the REVIEW, Crehore the first. In the *New Books* section, finally, Nichols' own *Laboratory Manual of Physics and Applied Electricity* was reviewed, briefly but favorably. Bedell and Merritt may have been busy, but so was Nichols; how he managed it all from Europe is not clear; he must have prepared most of it beforehand.

Also in Volume II was an article by G. W. Pierce at Texas on "Electric Strength" (he would become Harvard's big name in early radio circuit technology), a report on measurements with a

Cavendish balance to check the action of gravity on crystalline *versus* amorphous materials "at small distances," and a description of the new laboratory building at Adelbert College, which had been Western Reserve College, in Hudson, Ohio, before its move to Cleveland to ultimately become Case Western, where in a dingy basement room Michelson and Morley worked. There was a *Minor Contribution* from Sanford on his "electric photography" with two cuts reproducing photographs of coin features obtained with a charged coin in contact with the photographic plate. This contribution is followed by him again in a "Reply to Professor Carhart," who in Volume I criticizes a paper Sanford published in the *Philosophical Magazine* before the REVIEW birth, on the electrical conductivity of copper as affected by its surroundings (air, alcohol, kerosene). Carhart has found no effect. In "Reply to Professor Sanford," Carhart defends his position. Next in "Final Reply to Professor Carhart," Sanford finds Carhart's explanation of the discrepancy "entirely satisfactory." His only reason for "replying to Professor Carhart's (first) article was his assumption of superior accuracy in his work." Spirited argument, very professional, appears to wind up a draw. Books reviewed, rather better known than that of Nichols, were Raleigh's "Theory of Sound," Volume I, Cajori's *History of Mathematics*, and Preston's *Theory of Heat*. Issues of Volume II had two or three pages of book ads from Macmillan and at least one issue (Number 4 again) had an advertisement from Elmer G. Willyoung & Co., from Philadelphia, "Sole American agents for Nalder Bros. & Co., London," offering electrical instrumentation, with "Prof. Lodge's High Sensitivity Galvanometer" pictured.

For Volume II, 1895–1896, Bedell had a few problems. It is not clear how the editorial chores were divided between the editors but three letters to him indicate that he may have handled correspondence with contributors. Cornell's first Professor of Physics, Eli Blake, had died at Brown. A memoir was written by H. Appleton. Appleton was not pleased and wrote Bedell a curt letter of complaint; he had not received a copy of the REVIEW nor any sign that his writing was even published. (It was.) "Yet, free of charge, I expended both time and money in getting my copy ready ... your part of the task ought to be attended to ... ." A letter from J. Willard Gibbs informed Bedell that he "Can't undertake a REVIEW of Nernst"—in the Palmer translation. (It was reviewed by E. Buckingham of Bryn Mawr in Volume IV.) Michael Pupin tried writing Bedell but ... "I forgot to put on the stamp (2¢ red, Washington profile) and Uncle Sam returned it to me." He eventually got through; he liked Bedell's treatment of the transformer. No matter such problems; the year saw three well-known names published in Volume III: G.E.'s "wizard," Charles Steinmetz, writing

"Notes on the Theory of Oscillating Currents," K. Angstrom for "Photographically Registering the Infra-red", and R.A. Millikan (already) on "A Study of the Polarization of the Light Emitted by Incandescent Solid and Liquid Surfaces." He was not yet into charged droplets in electric fields; much more would be heard from him in REVIEW pages, almost to the end of World War II. A.G. Webster, of whom more will also be heard, in this third volume reviewed Hertz' book *Electric Waves, Being Researches on the Propagation of Electric Action with Finite Velocity through Space* and his *Die Principien der Mechanic in Neuen Zussammenhange dargestellt.*

It spite of a dearth of substantial articles submitted for publication, not all that were submitted were accepted. An archives letter from one Statefeldt asks that his paper, "A Modern Concept of Matter and Ether," be returned. It was apparently not accepted for publication. Presumably the referee in the case was one of the three editors. It is not known when the universal use of outside referees was put into practice (in 1901 Barnett is annoyed with an outside referee), but certainly it was in place by the time of Series II, when the Managing Editor was augmented by an Editorial Board. It is fortunate today that it is not a single editor or his board who alone has to decide on the suitability of papers. Statefeldt's paper was surely not unique of its type. Even (and perhaps more so) in these days we have "crackpot" writers and "philosopher-scientists" reporting on their work in private publications, and indeed, by members of the American Physical Society, in oral reports at meetings, their abstracts published in the Society *Bulletin.* If one is a member, anything goes. One must be careful in such labels and snickering; what is crackpot to one generation may be seen as unrecognized genius to another. Still, one T. B. Ford is probably not genius; he wrote Merritt concerning the "point of greatest favorable temperature for the production of electricity" (?!!), even invoking data on the great pyramid. Of more interest is another letter to Merritt: the well known optician, Brashear, writes that for $45 he can make for Merritt a three inch prism of rock salt. A dollar went further back then. There was a lot of communication between Merritt and Brashear, not connected with the REVIEW.

At the time, science was really stirring. The journal *Science,* which Bell had tried to rescue, made a new start under James Cattell, with Volume I of a new Series appearing in the first week of 1895. Published, as was the REVIEW, by Macmillan Company, it survived and has a solid reputation today as the organ of the A.A.A.S. A full-page ad in the PHYSICAL REVIEW (June 1901) lists an extensive Editorial Committee (including T. C. Mendenhall, for physics) and cites the need and predicts a bright future in a statement that bears quoting as indicative of growing appreciation and the needs of science:

> The past history of Science is a sufficient guarantee of its future usefulness. Such a journal is essential to the advance and proper recognition of the scientific work of each country, and in America where men of science are scattered over a great area, with no single center for personal intercourse, it is peculiarly needful. With the growth of science and scientific institutions in America, Science will occupy an even more important position than at present. It will continue to set a standard to the popular press in its treatment of scientific topics, to secure the general interest in science so essential to its material support, to enlarge the place of science in education and in life, and to demonstrate and increase the unity of science and the common interest of men of science.

It seems to have fairly well met the prospectus.

The PHYSICAL REVIEW faced more serious physics competition from another journal. Even as early as in the second year of the REVIEW, George Ellery Hale was writing Merritt about exchanging advertising for the PHYSICAL REVIEW and a planned *Astrophysical Journal*, commenting as well on his plans for a large observatory, clearly Yerkes (opened in 1894) "... still on paper." (He also inquires of galvanometers and bolometers; for the latter he finds the "smallest Swiss watch springs best.") In January 1896, two and a half years after the first REVIEW, the new magazine appears: Volume I, Number 1, of today's prestigious *Astrophysical Journal*, edited by Hale, then at Yerkes, and James Keeler of Allegheny Observatory. Whether this perturbed the conductors of the PHYSICAL REVIEW, or not, is not known. It might well have. The lead article was by Michelson on "Conditions which affect Spectrophotography of the Sun" (instrumental conditions in the spectroheliograph of Jansen, as perfected by Hale and Deslandres). There was a spectroscopic paper by Rowland and Tatnall and then one under Rowland's name alone—the famous table of solar spectrum wavelengths. The issue included a paper of E. E. Barnard with two of his pioneering and remarkable photographs of the Milky Way. A lot of spectroscopic work would appear in the Journal which could well have appeared in the REVIEW. Both Michelson and Rowland were on Hale's Board of Editors; Rowland never published in the REVIEW and Michelson only as three or four Physical Society meeting abstracts—and these presented only at meetings held at home in Chicago. The REVIEW seems never to have had a formal board of editors (naiveté?) until Series II of the journal came along. In contrast, right from the start, the *Astrophysical Journal* had an impressive editorial set up, the illustriousness of which likely gave the journal the air of maturity. The two Editors had four Assistant Editors: Ames of Hopkins, Crew of Northwestern (both physicists), Frost of Dartmouth and Campbell of Lick Observatory (astronomers). In addition there were ten Associate Editors: five foreigners (including physicist Cornu) and three Americans besides Rowland and Michelson: Hastings of Yale (physicist), Pickering of Harvard and Young of Princeton (both astronomers); Hale was clearly an operator and entrepreneur.

The papers in the *Astrophysical Journal* strike one as of more substance and as being more germane than those in the Cornell publication. The Journal was an obvious outgrowth of an earlier journal, *Astronomy and Astrophysics*, published at Carleton College. Volume I, Number 4, of the PHYSICAL REVIEW carries on the inside back cover as an advertisement, the index cover page for the December 1893 *Astronomy and Astrophysics*, edited by Wm. W. Payne and George E. Hale. The index includes a contribution by Lord Rayleigh, "The Theory of Stellar Scintillations." A similar back cover ad in REVIEW Volume III (Number 4) lists December 1894 papers and photos of E. E. Barnard, Percival Lowell ("Mars", naturally) and a four page paper by Hale, "The Astrophysical Journal."

Not only from astronomy was competition coming; chemistry also felt the need for a new journal, and this time it came from just next door to the REVIEW on the Cornell campus in the University's own Chemistry Department, where Wilder D. Bancroft and Joseph E. Trevor started the well known *Journal of Physical Chemistry*, with themselves as editors. As in the case of the PHYSICAL REVIEW, the editors were heavy contributors well supported by other Cornellians, the lead paper in Issue Number 1 (Oct., 1896) being by A. E. Taylor, a student of Bancroft's. The journal carried extensively abstracts of pertinent papers in other journals, many book reviews, and the cover pages (with indices) of the PHYSICAL REVIEW, *Astrophysical Journal*, and the philosophical *Monist Quarterly* of London, along with notice from the "Chemical Department of Cornell University" of courses of instruction offered in physical chemistry for the academic year ahead, in the list of eight, seven being given under the auspices of the two editors.

The pair of editors of the physical chemistry journal was a curious duo. They were both independently wealthy, each donating his service to the University. Trevor later apparently suffered some reverses; an archives letter to the University President from a Department Chairman suggests some remuneration would be appropriate for him and welcome. He was a thin wisp of a man, smoker of strong cigars, a shadow, rather much the recluse, although he did play in a quartet with other faculty members on Saturday nights in the Physics laboratory. It is strange that this almost painfully shy man got involved with his co-editor, who was very much the opposite. Bancroft enjoyed being a character and could frequently be seen wending his way up the hill to the Chemistry building, clad in his scarlet academic gown. He came from a famous Boston family, one of his forebears being our first Secretary of the Navy. He was wealthy enough that he was referred to on campus as a "millionaire"—something in those days. He never collected his salary, was the first on campus to own his own automobile, which he never drove. A good thing it is said, for he broke everything he touched. It was his money which kept the *Journal of Physical Chemistry* solvent until it was taken over by the American Chemical Society. No Cornell subsidy on this one. Trevor was a classical thermodynamicist and he eventually dropped out of the editorial work and left the journal to Bancroft; the latter had little use for formal thermodynamics and his journal was a medium through which he could allow the subject to be disparaged by unfriendly and biased writers. He favored the descriptive mode and heartily endorsed Gibbs' phase rule in that format. The difference between Bancroft and Trevor probably led the latter to seek his base, quietly, in the Physics Department. Bancroft held the editorship until 1932, when it was taken from him not easily, in actions by the Chemical Society, the Chemical Foundation and the new American Institute of Physics, egged on by young physical chemists led by Urey at Columbia. John W. Servos, in a long article in *Isis* (1982), relates what went on in describing Bancroft's years with his journal.

As in the case of the *Astrophysical Journal*, here again, in the *Journal of Physical Chemistry*, there appeared many papers that would have been appropriate for our physics journal. But science was growing; in spite of the new competition, the REVIEW in Volume V, July 1897, became a monthly, a new volume appearing every six months after at least five issues per volume.

Further evidence of the growth of science generally was the appearance a bit later, in 1898, of the very useful and much needed (ever so much more so today than then) *Science Abstracts*, which for many years came regularly to members of the American Physical Society as subscribers to the PHYSICAL REVIEW. (No longer.) A letter dated March 8, 1898, sent to the Editors of the REVIEW from the Institution of Electrical Engineers in Great Britain, announces the debut of the *Abstracts* and would like to engage in an exchange of journals so that they may abstract articles from the Cornell publication. Abstract Number 1 in January of Volume 1 is of a paper on "The Process of Solidification" by G. Tammann, which appeared in *Annal. Phys. Chem.*, **62.2**, pp. 280-299. The list of journals from which digests were printed included the PHYSICAL REVIEW, the *Auto–Motor and Horseless Vehicle Journal, Nuovo Cimento, Comptes Rendus, Nature,* the *Astrophysical Journal*, the *Railroad Gazette*, and the *Street Railway Review*, among others of serious and/or quaint title. The asterisk before the names of some publications listed indicated that that journal was only occasionally abstracted. In spite of that disclaimer and no asterisk before its name, there was in the first issue nothing from the PHYSICAL REVIEW but much from the *Astrophysical Journal* and the *Journal of Physical Chemistry*. Someone had been minding the store in astronomy and chemistry, or perhaps more to the point, the abstractors found more meat in them than in our REVIEW. Abstract Number 30, in fact, was from the chemistry journal, of a paper by editor Bancroft. Interestingly, Number 23 abstracts "Interferential Spectroscopy" by A. Perot and C. Fabry. It was not until the next issue that the first PHYSICAL REVIEW paper was abstracted: Number 112, "Arc Spectra" by A. L. Foley, work done at Cornell under Nichols. The next from the REVIEW was in the next *Abstracts* issue by another Cornellian: Number 232 by N. E. Dorsey on "Surface Tension". Number 105, in February, abstracts "Atomic Hypothesis" by L. Boltzmann, *Anal. Phys. Chem.*, responding to one Volkmann's negative argument concerning the discreet atomistics concept and differential equations, Boltzmann pointing out that the latter concept is not so unlike the former involving as it does, discreet finite space elements, etc. This was followed by some abstracts (Becquerel, Cornu, Cotton) on the new effect of Zeeman. Abstract Number 1034 is a three and a half page item by W. C. Roentgen on "Roentgen Rays." There were many papers on rays: Canal Rays, Cathode Rays, Roentgen Rays, and one on Uranium Rays ("... and Condensation of Water Vapor") by one C.T.R. Wilson, Number 1033, in fact. Familiar names all. There were papers on Dynamos, Telephony, and Telegraphy, these abstracted from such as the *Electrical Review*, *L'Undustrie Electrique*, etc. Late in the year from *Comptes Rendus* came an abstract of a paper by Slodowska–Curie reporting something in pitch blende more active than even uranium: "... it seems as if these (uranium bearing) materials contained some element more powerful than uranium (the metal) itself." In the next volume, abstract Number 35 is of a paper by P. Curie and Mme. Slodowska–Curie announcing the isolation of polonium and in Number 664, again with her husband, and H. Bemont, the separation of "radium." Nuclear physics was on the horizon.

Earlier, however, in late 1895 the discovery occurred in Vienna from which we are saying that modern physics took off. (We discount the troubling ether drift experiment, which, while coming eight years earlier, did not really influence the early growth of modern physics, remaining a puzzle until it was seen as a natural consequence of Einstein's new relativity nearly twenty years later.) This was Roentgen's discovery of X-radiation. As we have noted, people at Cornell got right into the action. Moler was taking X-ray pictures of Mrs. Ryan's hand, Nichols was lecturing over at Albany on the subject. Yet Volume IV of the REVIEW (1896–1897, the year of Becquerel, of Zeeman) gave no inkling that something new and important had been discovered. There were two short *Notes* by Robt. Wm. Wood (like Millikan, to become a long term contributor), one on the Doppler effect (acoustic) and one on a demonstration experiment showing orbit motion; steel balls rolling over a smoked glass plate under which is the pole of a magnet which deflects the "particles" away from otherwise straight line trajectory, into curved orbits around it, "... useful in making more cheerful that portion of the (physics) course normally rather destitute of pyrotechnics" in the words of this well known pyrotechnician. (Recall his strolling in Baltimore slums after a rain shower, spitting into puddles along his way, flicking a pellet of sodium at the same time at the same target, to baffle and unsettle the casual observer; his mounting and setting to spin a heavy gyroscope in a suitcase handed over to a porter and making their way through a crowded station; his disposing of n-radiation by stealing the prism from the apparatus of the "discoverer" with no apparent diminution of the intensity; his children's book on "Telling the Birds from the Flowers.") The orbit paper includes a typical free sketch (of the apparatus) with which his later text (*Physical Optics*) is replete.

The fourth REVIEW volume, further, carries a contributed paper by Morley (no more either drift) with Rogers (of Volume I, Number 1), on the thermal expansion of metals using the interferometer (naturally enough); an experimental paper on the "Velocity of Electric Waves" by Saunders under Webster up at Clark University, and a paper on "Heat Rays of Great Wavelength" (reststrahlen) by Rubens and Nichols (E. F. this time), preceded by one of Nichols, solo, describing his use of the Crookes radiometer in infra-red measurements on quartz and in the reststrahlen region, both papers of rather landmark status. But no X-rays. The Saunders' paper seems fairly remarkable; in a long parallel wires system, excited by a spark, resonance is observed by sparking across the gap at the far end, the period of which is obtained in means reminiscent of Michelson, by observation with a rotating mirror. In the Cornell archives, Nichols (E. F.) writes to Nichols (E. L.) concerning the cuts of his article with Rubens and writes a letter to Merritt including a sketch of Boltzmann's interference mirrors used for making infra-red wavelength measurements, a technique re-discovered in the mid-twentieth century resurgence of infra-red investigation.

REVIEW Volume V (1897) recognized the X-ray discovery. R. W. Wood has a paper on a new cathode discharge for intense production of X-rays; he is looking for diffraction of the radiation in possible fringes from a narrow slit; there is some indication of such. A paper on the discharge of electrified bodies with X-rays by C. D. Child, known better for his equation governing space charge limited current flow, reviews experiments and notes the effect of air density in an experiment of his own on the discharge by Becquerel rays from uranium. Fernando Sanford is present again, writing with Lillian Ray "On a Possible Change in Weight in Chemical Reactions" reported by Landolt a year earlier in the *Zeitschrift für Physik*. They find it improbable in their measurements. There is a review of books already written on X-rays; those on radioactivity would come along. There is another death to report; the frontispiece portrait is of Alfred Mayer of Lehigh, worker in acoustics. And no effect is seen in a gravitational permeability experiment carried out with a Cavendish balance at Wisconsin, duly reported in a paper of the volume; so there is some basic physics going on and carried by the REVIEW; it is not all trivia. Still, J. J. Thompson did not report his 1897 discovery of the electron in our pages. The best physics was clearly being done across the Atlantic.

At the bottom of the last page of the July issue of 1897, Volume V, Number 1, there appears this note:

Notice: Change from Annual to Semi-annual Volumes.

Beginning with the present number, two volumes of the PHYSICAL REVIEW will be published annually. These volumes will begin in July and January respectively and will contain at least five numbers. The past volumes of the PHYSICAL REVIEW (I-IV) are annual volumes, each containing six bi-monthly numbers, beginning with the July-August number, 1893.

No mention of advertising but with the next issue, August, the Cornell basement holdings of "originals" appear with covers and with advertising, both at the front and back of each issue. The inside of the front cover carries the above information and notes further that:

The price of subscription is two dollars and fifty cents a volume (five dollars a year) or fifty cents a number.

Subscriptions are to be sent to the publisher, Macmillan and Company, 66 Fifth Avenue, New York; Macmillan & Co., London; or to Messrs. Mayer and Mueller, Berlin. Back issues could be had from the publishers at the former subscription price, three dollars per volume.

> *Correspondence related to contributions (manuscripts, that is)*
> *should be addressed to the editors at Ithaca, New York.*
> *Manuscripts intended for publication in the* PHYSICAL REVIEW
> *must be communicated by the author; when publication in other*
> *journals is contemplated, notice to that effect should be given.*
> *The authors of original articles published in the* REVIEW *will*
> *receive one hundred separate copies in covers, for which no charge*
> *will be made; additional copies, when ordered in advance may be*
> *obtained at cost.*

The advertising at this time was not extensive and pretty much devoted to publishing. Macmillan obviously had texts for sale, including: *The Outlines of Physics—an Elementary Text-Book* by Edward L. Nichols; others by Nichols and Cornell co-authors; *Alternating Currents and Alternating Current Machinery* by Jackson and Jackson; *The Principles of the Transformer* by Frederick Bedell, Ph.D., which comes in for glowing comment, including that from *Automotor* ("... author has entirely succeeded ... work is invaluable..."); *Theory of Electricity and Magnetism* by A. G. Webster. The Open Court Publishing Company was pushing Ernest Mach's works, *The Analysis of Sensations, The Science of Mechanics,* and *Popular Science Lectures*. *The Journal of Physical Chemistry* had a full page advertisement listing Recent Articles, six of which were by editor Bancroft. The *Astrophysical Journal* had a page giving "Contents of Recent Numbers," including some noted authors; H. Kayser on the spectrum of hydrogen, Hale himself in a series of papers on Yerkes Observatory, Evershed on the darkness of sun spots, P. Zeeman on the influence of magnetism on the nature of light emitted by a substance, a follow-up article by Michelson on "Radiation in a Magnetic Field," and instrumental papers by F.L.O. Wadsworth, all papers seemingly of more interest than things in the PHYSICAL REVIEW. Macmillan was advertising *Nature* and *Science*, offering them both, postage prepaid, for $10. There was little in the way of apparatus: Peerless Typewriter ("True Merit Wins Success") Co. of Ithaca; Nalder Bros. & Co. of London ("No agents in the United States"; Willyoung and Nalder must have had a falling out after Volume II, Number 4) advertising the Trotter Bar photometer @ $32 ("Much easier to read than either the shadow or grease spot instrument.") and an Ayrton–Mather Universal Shunt, New Pattern @ $16. A bit out of line was this quarter page from Rumford Chemical Works in Providence: A TONIC—for Brain workers, the Weak and Debilitated; "Horsford's Acid Phosphate—without exception the Best Remedy for relieving Mental and Nervous Exhaustion; and where the system has become debilitated by disease, it acts as a general tonic and vitalizer, affording sustenance to brain and body."

(Beware of Substitutes and Imitations.) A year later, Western Electrical Instrument Co. of Newark (offering Weston portable instruments), Wagner Electric, General Electric, and Westinghouse had joined Nalder. Macmillan had a lot more books: in Mechanics they included *An Elementary Treatise on Theoretical Mechanics* by Alexander Ziwet (Michigan) as well as *The Mechanics of Hoisting Machinery—including Accumulators, Excavators, and Pile Drivers.* And the REVIEW itself was offering "PORTRAITS"—to satisfy a demand—issue a limited number—scientific men—appeared as frontispieces in the PHYSICAL REVIEW. Printed in fine photogravure, plate paper, suitable for framing—very low price of 25¢ each. Ready were: Tyndall, Hertz, Helmholtz, Kundt, and Mayer. By the time of Volume XII (1901) Rogers, Wiedemann, Bunsen, and Preston were also available. And advertising was up: Gaertner with Michelson's interferometer (in a well recognized version to an old physicist), the full page ads for *Science*, for Weston Electric Instruments, General Electric (a full page offering "pocket" ammeters) and an impressive page of Queen & Co. (Philadelphia), purveyors of physical apparatus, electrical testing instruments and self-regulating X-ray tubes, with a picture of their Meter Spark Induction Coil ("produces heavy 46" spark, the largest of the kind ever made"). More modest at the back with the book ads is Ziegler Electric Co.'s (Boston) Ziegler hand dynamo (4 amps at 50 volts), the armature driven by a hand crank like an old ice cream freezer, geared up undoubtedly; as tiring to a 100 watt operator as if he were winding up a freezer of cream.

If the important discoveries were being made in Europe, American physicists seemed generally slow in getting in on the action. In 1896 Zeeman discovered the effect of a magnetic field on the radiation of a bright line source within it and Lorentz had provided the beautiful classical explanation of the phenomenon. But it was not until two years later in Volume IX of the PHYSICAL REVIEW that Shedd, then at Wisconsin, reported using the suggestion that Michelson made in the *Astrophysical Journal* that the interferometer was the ideal instrument to use in study of the effect. Frontispieces in the volume were of Wiedemann and Bunsen, E.L.N. again writing the memoirs. Papers included one on "An Investigation of the Magnetic Qualities of Building Brick" (!), the authors worrying about magnetic effects of thirty brick samples used in the construction of physics laboratories; and a paper from Cornell by one L. W. Hartman on "The Photometric Study of Acetylene and Hydrogen Mixtures Burned in Air"; radiation standard sources and photometry were big in Cornell physics. More important in this early period was a twelve-page resume as a *Minor Contribution* in Volume VI on experiments with Becquerel rays by O. M. Stewart at Cornell (more arm twisting?). The same volume also included a paper which got Michelson about as close to publishing a real paper in the REVIEW as he would come. This was a paper "On the Electrical Resistance of Thin Films" by Isabelle Stone, summarizing a Ph.D. thesis she had done under the great man's direction at Chicago. And there was an interesting paper in the next volume (VII) by Merritt on "The Magnetic Deflection of Reflected (and Transmitted) Kathode Rays." Unfortunately the technique was inadequate; he was twenty-five years too early. Also in Volume VII was a rather good paper by Henry Eddy, our man Morley, and Dayton C. Miller, who, using an interferometer at Case Western again, find no change as great as a part in 60 million in the velocity of plane polarized light propagating through carbon disulphide, a container of which upon the application of a magnetic field shows as much as a full half turn in the plane of polarization of the light passing through it. And in an instrumental paper, LeConte Stevens describes a mechanical harmonic analyzer for wave form analysis providing coefficients for up to eleven harmonics, this also in Volume VII.

The citation of particular papers from successive volumes of the REVIEW as we have been doing in the foregoing, and in what is yet to come, may give an overview of the journal and its development over the years, but it does not serve to tell what was going on in the editorial office. The documentation of some of these early years found in the Cornell Physics' attic and now in proper Cornell archives, serves better to illustrate the less formal side of the endeavor. Much of the material is simply routine, dull, business with Macmillan and New Era Printing. There are acknowledgments of receipt of manuscripts, thank you. There is personal correspondence, not especially germane to the REVIEW but some of it interesting to a physicist. There are books of receipts of payments made on advertising, carbons of bills paid, inquiries about physics, some correspondence relative to teaching activities, sundry other trivia, some of it perhaps of interest.

There are, for example, three serial post cards from England by Larmor which concern his coming memoir on Fitzgerald. And from across the way at Cornell, J. E. Trevor sends Bedell several deeply apologetic memos about his delay in turning in the memorial he has agreed to do on J. Willard Gibbs: he is presently swamped but in the next week or so—; an academic recess is nigh and he should find the time—; etc. But he gets it done and his long, considered published statement on the man and his work stands with Larmor's on Fitzgerald. From Leiden, H. A. Lorentz writes Merritt: "I thank you most heartily for sending me Ns. CXXVII and CLXV (Oct. 1906, Dec. 1909) of The PHYSICAL REVIEW which I duly received." One guesses that he was interested in the reflection of radio waves by Nichols' (E. F.) screen mesh in CXXVII and in two articles on electric and magnetic double refraction in Liquids in CLXV. From Manchester, Rutherford is missing some copies: " ... much obliged if they can be sent. We are all well and flourishing but busy as usual." Understandably. "With kind regards, ... " So with W. H. Bragg; he has not received a certain copy and he would like to submit a *Note* also. G. B. Pegram, at the U.S. Coast and Geodetic Survey has sent Merritt some amber, sorry to have missed Rutherford; he's into radio-activity himself. R. W. Wood asks Merritt for a loan of some Crookes' tubes and asks advice on a spectrometer—he has $300 to $350 to spend; should it be a German instrument or one from Societe Genevoise? Michelson writes Nichols about Cornell's Physics Department so he can make a case before the Chicago Trustees. His is beautiful calligraphy. So too that of G. E.'s giant, Charles Steinmetz; would that we all wrote so attractively. LeConte Stevens was runner up in style. Other handwriting is deplorable, almost unreadable; why some individuals feel they must sign communications in an undecipherable scrawl is strange—psychological problems no doubt. There is a letter from Morley in Florence regarding his paper with Rogers on the expansion of metals using interferometry, and in another letter later he declines regretfully to do the memorial on Rogers; he's had his own family deaths of late and is not well.

There is a letter from Robert A. Millikan; would it be possible to publish an article on the polarization of light from incandescent solids and liquids? It was. A letter later to Merritt from him expresses pleasure and surprise that it was already out; he had expected to see proof again but the cuts were better than what he had sent in. Please send ten reprints to Ogden Rood (our friend in Cornell's history, above) and the remainder to A. F. Millikan in Oak Park. It was well to have made here a good impression; Millikan would be a constant and prolific contributor in years to come, which surely did not hurt the REVIEW's reputation. In 1900 E. Hall at Harvard wonders about the possibility of a paper he can do; he demonstrates to his students the meniscus disappearance at the critical temperature in a tube containing CO_2; every year he is fearful of an explosion but he has it now encapsulated, under control, and the problem safely in hand. Would the editors be interested? Yes, indeed, and it was published. Before going back to Britain, Rutherford from McGill asks about publication of his "Discharge from Glowing Platinum and the Velocity of the Ions." No problem. Lyman J. Briggs, then in the Department of Agriculture (Soils Division; the Bureau of Standards Directorship yet to come) in Washington, writes that he is employed as a physicist and offers to do the yearly indexing in exchange for a free subscription. The editors apparently took him up on it; a letter or two from him encloses the index for the year. A few years later, he requests discharge, the Physical Society having taken the journal on.

The "editorial office" was not free from the reception of complaints. In a letter dated Dec. 24, 1900, one Judd, from Cincinnati, in the spirit of Christmas writes Merritt "... certainly can not consent to being in any way connected with these plates that you propose printing ... would be no point to the paper and the method would appear without merit (not Merritt)." Perhaps both points were well taken; the name Judd does not appear in the cumulative index, Series I. In 1901 Child writes to Merritt from over at Hamilton. He has the galley proof of his article; "In the equations the typesetter seems to have put in almost any letter that came handy." Child has the original and the errors there are not his fault. "A person that can not tell χ from X, or a from A, or dv/dv from dV/dX had better quit." And Barnett in 1901 complains that the referee at the *Philosophical Magazine*, who did not find all well in Barnett's submission on the Cavendish experiment and the inverse square law, "doesn't understand it at all." But Barnett had tried to make things clearer in the rewrite. It was published in Volume XV next to his *Note* on Gauss' Theorem. Bevier, in one of his numerous vowel investigations, was annoyed after his first article. He had sent the paper in with a formula incorporated for his "energy" plot: $E \propto a^2 n^2$. In the proof he objected that the proportionality sign looked like an alpha; fix it up. They did; his annoyance lay in how it was done: E varries (sic!) as $a^2 n^2$; after all, he could spell. And there was Appleton's extreme displeasure with Bedell's not acknowledging his memoir on Eli Blake.

Mail seems to have been a problem—to both the REVIEW and its correspondents—and not only in missing copies. We have noted Pupin's absent-mindedly posting his letter without stamp. R. W. Wood got into the same trouble; he was ill, out of stamps, and trusted an assistant to mail his communication to Bedell, which the fellow did—by just dropping it in the box without stamps. Bedell had to recover that one from the dead letter office; Wood apologized for his trouble. And a letter from Alexander Ziwet of Michigan tells that a copy of the PHYSICAL REVIEW containing the notice of his "Mechanics III" had just reached him in Ann Arbor by way of London, England. "Thanking you for your kindness in sending the copy," he writes, "I beg leave to inform you that I am not an Englishman, and remain, Faithfully yours, etc."

Book reviews earned correspondence. We have noted some. C. R. Mann, in optics at Chicago, has made his review of *Edser's Optics* short, since "I do not think the book's such a *chef d'oevre* as a text to warrant larger notice." There may have been slight conflict of interest here with his own *Manual of Advanced Optics* published in the same year. Still, the review, short enough, is not all that adverse. He made no note of Edser's characteristic figures, black background with white lines thereon, the negative of the usual. A letter from A. G. Webster reports that he is letting Ames at Hopkins do the review of Lord Kelvin's lectures; Ames was agreeable and is thoroughly familiar with the subjects. The editors seem to have been fairly relaxed about such goings on. Southall wonders if he can have a review of his book on optics; unfortunately by then the editors had stopped the inclusion of extended book reviews, but they would print notice of the book's being published, which they did under *New Books*. And a letter from E. B. Frost, Dartmouth, agrees to do a review of one William Ford Stanley's *Notes on the Nebular Theory*, Frost's opinion given in Volume III of the Review. He may have felt that he'd been had. We quote pieces of the Review:

> ... *contains speculations upon the Nebular Theory of a gentleman not a professional astronomer. (Stanley, however given to much thought on the subject.) Never-the-less, it is difficult to regard the work as a positive contribution to our knowledge of cosmogony. ... many views advanced can not be harmonized with the laws of mechanics ... style not lucid ... language so clogged with words it is difficult to grasp an idea presented. Example, not particularly conspicuous; "In this case, with proportional time condensation under increasing amount of tangential impulse due to centralized condensation into gravitation which produces the law of orbit, the distances of the planets from the sun and their separate masses would be symmetrical proportional, in accordance with the pull of gravitation and the tangential momentum of the amount of condensed matter. Etc, etc.*
> *The typographical appearance of the book is excellent.*

There are not many lapses of this sort in the Review compendium over the years. This writer would be hard pressed to prove that of a great many theoretical papers of recent times; but at least the verbiage today makes sense. The really wild theories these days are consigned to APS meetings, where seemingly anything goes if the author is a member of the Society. Dr. Charles Brush, an experimenter in Cleveland, was not quite of this ilk, but close enough. He will have some abstracts in the Review of some of his surprising gravitational acceleration experiments, which occasioned some precision measurements, here and abroad on the relative gravitational acceleration experienced by various materials. He actually had a paper published in the *Philosophical Magazine*, as he pointed out in submitting a paper to the Physical Review on "Etherion—a New Gas," which he also submitted to Cattell's *Science* and to other journals. Cattell wrote Nichols about it; his magazine will publish it "this week or next. It is certainly an extraordinary thing. I scarcely know what to think of it." Understandable enough on reading in *Science Abstracts* for 1898, Brush's Abstract, Number 325: "From the high heat conductivity of high vacuum, it is concluded that a gas of insignificant molecular weight but of high specific heat and molecular velocity exists in glass and other solids." In a later abstract, Crookes believes Brush is simply tangling with "aqueous vapor."

While one might take Brush's conclusions and results with a large grain of salt, at least his prose was coherent. Not so that of Stanley above; nor that of T. B. Ford of Middletown, N.Y., noted also earlier. He wanted publication of his "Point of Greatest Favorable Temperature for the Natural Production of Electricity." Merritt wrote back that he did not understand what was meant by the point of greatest favorable temperature or the principle of natural production. Ford replies in a long handwritten letter which is almost pure gibberish; a measurement on the base of the great pyramid enters. A following letter by Merritt advises Ford to familiarize himself with modern views and terminology; and thanks him for copies of "Conglomerate" sent with Ford's letter and explanation, "... interested me greatly." Ford will be sent a sample of the REVIEW with the next issue. A. K. Bartlett, astronomer, submits a paper on "The Wet and Dry Moon;" he has written on the subject for twenty years and knows it cold. His paper is not found in the REVIEW cumulative Index of 1920. Nor is one to be found of McLennen's of Brooklyn (Iowa); he had sent in a manuscript, along with favorable comment on his earlier "Cosmic Evolution," expressed by the *Boston Advertiser* and the *Chicago Inter-Ocean*, and such. One is sympathetic with McLennen. He has sent postage to pay for the manuscript return if it is rejected. He "realizes that it is not all improbable that I am wrong." He would appreciate being led to the right for the "sole object is the ascertainment of the truth." He would appreciate reasons for the possible (probable?) rejection, "especially whether or not you think my criticisms of Herschel and Newton erroneous." Nor did S. T. Preston of Germany have the courage of his convictions; there are three letters concerning *his* "Cosmological Theories: an Inversion of Ideas as to the Structure of the Universe"; he too sent in postage with the manuscript for its return. Both writers inquire later as to what has become of their entries. One other cosmologist, Robert Stevenson (not Louis) can send in his manuscript if they'd like, on Weight, Energy, Resilience (and elasticity and gravity); he will prove a fallacy in the 2nd law of motion; the true cause of the various manifestations is Motion; if the editors prefer, the journal can use his "nom de plume 'Elasticus'." No Stevenson, no Elasticus in the Index. One last such from Eastbourne, England, comes a manuscript "Nature's Voice"; "... opens up a new view of the universe—most magnificent. What I earnestly plead is that you will carefully master the subject and see the glories of universal nature as I see them ... etc." Later comes a follow-up letter acknowledging its return with polite thanks for the trouble. Some such offerings draw sympathy, others, ire. A penciled letter on lined paper, the writer recognizing his lack of education, his desire to further understanding, is there any worth in it, take what is useful, sorry to trouble you, etc., brings a far different reaction than does a letter of arrogance, this is the answer to everything, how can you be so stupid as to not see the truth in the effort, show me where I'm wrong, etc. The wish to understand the physical universe seems to come at all levels of erudition. For the most part they are not charletons. Hardly so the operator in Atlanta who is hawking stock around the town to support a gold mining scheme on which Engineering Professor Quick wants Merritt's opinion; Quick has not the physics insight to judge adequately but has his suspicions. Two electrodes on either side of a mountain full of gold are driven down to the water table and connected to the secondary of a giant Ruhmkorff coil interrupted at 100,000 cycles per minute, gold vibrating (!) at a multiple of that frequency. Current flows through the mountain, plating gold out on one electrode or the other. Quick sizes up the fellow as a "rogue," probably leaving town after bilking a few suckers, accumulating gold we might add in a somewhat more straightforward and easy manner.

There are a lot of letters concerning exchange of journals with other journals, some more appropriate than others: *Western Field—The Sportsman's Magazine of the West*—"would appreciate an exchange of publications." Request for an exchange comes from the *Journal of Arya Patrika*, "the most intelligent religious society in India"; paramount importance to you, etc. Meyer Brothers Druggist sends in a copy for exchange. Not exactly an exchange is this from Lahore: "—a pleasure to inform you that the Literary Department of our firm wishes to contribute articles on Indian life and thought to your valuable and esteemed journal." Send them a copy to enable their forming an idea of the lines on which it is conducted. There were of course many sensible exchanges: the *Astrophysical Journal* (a letter or so from Hale), *Science Abstracts* (the letter announcing its publication), the *Journal of Physical Chemistry* (letter from J. E. Trevor), the *American Journal of Science* (letter from E. S. Dana) among others. On the other hand, Johann Ambrosius Barth, publisher in Germany, does not wish to exchange the PHYSICAL REVIEW with Wiedemann's *Annalen der Physik und Chemie*, which he publishes. Rather he prefers simply to subscribe to it, since the *Annalen* costs (36 mks) three times as much as the REVIEW. Sounds to be a pretty close operation. There is a lot of correspondence concerning such exchanges.

As we have noted, bound copies of Series I of our journal held by the Cornell Library contain no advertising pages. Yet the archives from these early days is loaded with letters on advertising and copies of receipts to advertisers. It was something of a revelation then to look at the Physics Department holdings of back copies in the original (after Volume V) to find extensive advertising; clearly the library did not bind up the pages of advertising. With Series II, advertising pages (some?) are bound in the Cornell holdings. The editors worked a lot of deals to the advantage of the Cornell Physics Department, offering advertising in exchange for good prices on equipment or apparatus, some acceptable to the advertiser, some not. Gaertner accepts the offer of "one half a page for the year and $50 each for one of our assortments No. 3." General Electric does better; accepts one full page for a year in exchange of $250 for a big selection of meters totaling $885 retail. The Caligraph Typewriter ("It Outlasts them All") Company would be willing "to take 25% of the price of our #4 Caligraph in advertising in your journal." The company "could not agree to the proposition you make to take out the entire value of the machine in advertising." Other typewriter deals are proposed: to Remington, Dougherty, Smith Premier, Peerless (made in Ithaca, handy enough); Peerless may have acceded, for a year later Chemistry Professor Dennis to Nichols "encloses a check for $60 to pay for the Peerless Typewriter," so it may not have been suitable REVIEW equipment. What the replacement was is undetermined. From another advertiser comes judgment of his ad sheet proof; it is okay if the cut of the advertised offering is turned right side up. And from Camillo Olivetti, Ivrea, Italy, comes this in 1901: "With astonishment I have seen on the last number of the PHYSICAL REVIEW one advertisement of mine, although I have written that I did not want to go on in this advertisement which has not proved to be a paying one. Should you want to occupy the page and go on advertising my products free, I beg to change the form of the advertisement, being ridiculous now to invite people to see my stand at the Paris Exposition." He had earlier proposed swapping some galvanometers for one half page for a year, reasonable enough.

The ad states:

> *Professors and Assistants in electrical laboratories don't like to give students fine galvanometers because such instruments are liable to be spoiled.*
>
> **They are perfectly right.**

Students don't like to work with cheaply made instruments because they are not correct and are troublesome to handle. They are perfectly right. Use Olivetti's Reflecting Galvanometers and you will have high grade, up to date instruments which don't fear hard service.

A cut shows his ware, one or two of them still extant as museum pieces in the Cornell Physics Department. From Germany came an urgent wire to Bedell:

Send immedy border and nameplate our ad PHYSICAL REVIEW *to Edw. Barton Illinois Water Sup assn urbana Ills same will be returned.*

Siemans and Halske

(In the first decade of the century, the Illinois State Water Survey, in Urbana, probably had better scientists in it than did the University Physics Department. While now vastly overshadowed by its illustrious neighbor, it is still a viable enterprise.) One supposes that what was to be inside the Illinois border was something different than what the REVIEW was running. The back side of the Western Union familiar yellow message sheet was filled with company propaganda: 1,000,000 miles of wire, 24,000 offices, 7 Atlantic cables and an engraving of a giant telegraph pole and cross arms bearing hundreds of wires, alongside of a real small one with few wires, emblematic of the sizes of Western Union and its competition. Ezra Cornell would have been pleased; but times have changed.

Even after its first decade, the REVIEW was not as well known as, say, the *Saturday Evening Post*. Professor Moler (unassociated with the journal) had a letter from the Western Electric Company in Chicago asking "... is a journal of physics formerly published at Cornell University still published?" And from the Los Angeles Aqueduct Authority a letter from an Engineer to Merritt in response to some sort of note from the latter; he notices that Merritt has responded on stationery under the masthead of a journal called the PHYSICAL REVIEW; sounds to be something he should be subscribing to; sign him up. One has to admit that there were more articles on viscosity of water, of freezing characteristics, etc. than there were on quantum mechanics back then.

Late in the decade the American Physical Society enters into some of the correspondence. At Webster's instigation, the idea of forming such a society has been broached at last. LeConte Stevens, now moved to Washington and Lee, writes in his fine hand in support of a Physical Society; he has anticipated such for a number of years and doesn't worry too much about the A.A.A.S. Section B; physicists must demand, and they do, a separate society. Franklin, by now at Lehigh, is in agreement that we should get a Physical Society going, is pleased with the REVIEW but thinks the business of authors having duplicate publications is not good. (And he advances an interesting notion in physics to Merritt: can't we measure the velocity of cathode rays by the light they give off (?) as a "bow wave, conical in its propagation about the direction of motion, and polarized?" Not that such light was known but he's not far off, considering Cerenkov decades later.)

Franklin's point about the double publication of papers had bothered others from the very first when McGregor indicated he had already submitted a requested paper to another journal and could not well do it for the REVIEW also. In January 1901, Carhart submitted his paper on the EMF of the Clark Cell although it would be published in the February *Electro-Chemist and Metallurgist*. But he saw no problem with that; there could be no objection to simultaneous publication, arguing that the *Philosophical Magazine* frequently did it. Shedd asks for blocks of a few of the cuts in his interferometer *versus* Zeeman effect paper; Colorado College wants tracts from some of its people and he is to contribute his bit. Somewhat different. No less a person than Ernest Nichols has sent the manuscript on the radiation pressure paper to the *Astrophysical Journal*; it is difficult for him to make changes preventing strong resemblance between it and what was sent the REVIEW: "... the paper is so intimately Physical and Astronomical in its leaning that the confusion is a natural one, aside from the peculiar embarrassment under which I am placed." He is presumably thinking of his close tie to the Cornell enterprise. But a couple of years after the founding of the REVIEW, in a letter to L. T. More, Merritt is pleased to have his manuscript; it will go in the November issue if it does not appear elsewhere first. There is no regular understanding with the *Philosophical Magazine* in regard to simultaneous publication, although neither they nor the REVIEW object to *simultaneous* appearance of an article in both magazines. If it comes to the REVIEW after the *Philosophical Magazine* does it, it will not be published by the REVIEW. It is not clear whether the criterion applied as between journals in the same country, as concerned Ernest Nichols. Officially, a policy came to be given on the inside of the front covers after the journal became a monthly, as has been cited. In any event it never became a real problem. Perhaps today, at the near end of the twentieth century, the most serious transgression of the like involves the PHYSICAL REVIEW and its offspring, the PHYSICAL REVIEW *Letters*. A person makes a red-hot discovery; he or she reports it to the *Letters* (perhaps preceded by oral report at a Physical Society meeting, as in days of yore) and then reports it again, in greater detail, admittedly, to the REVIEW. Since the intent of the *Letters* is largely one of quick announcement of important development and that of the REVIEW is that of full disclosure and recording for posterity, no serious breach of publishing decorum is seen; the practice in fact has been taken up by other journals, both here and abroad.

Like Franklin's conical "bow wave" of radiation suggestion, some of the other correspondence is perhaps of interest, although not pertaining particularly to the PHYSICAL REVIEW. Wood writes wondering of Merritt's plans for work in fluorescence; he wants to avoid any duplication; he has been getting some interesting results in the continuation of his work on sodium vapor by photographing the fluorescent spectrum when "stirred up" by monochromatic light of varying wavelength; he is in to his well-known work on resonance radiation. And he playfully chides Merritt over something he had sent in: "... that was not just one short paper—there were three; so each is only one third of a short paper." Wood worries that he has heard from someone critical of his steel ball orbits around a magnet pole; it wasn't an inverse square law field. But while it may have been implied, he never out-and-out said anything about the field dependence with distance. And in another letter on his intense X-ray source, he is concerned that Nicola Tesla has essentially the same idea, discovered independently, or he learned of Wood's by the grapevine. Anyway, Wood is satisfied that he has the priority. Foley inquires of what Merritt thinks is the basis of the well-known phenomenon that the negative leg of a lamp filament wore away a lot faster than the positive leg; the Edison effect may be involved. There are a number of letters from G.E. (division in Harrison, N.J.) about some special lamps Merritt wanted fabricated for study of the effect. Hartman, from Kansas, could not recall how Merritt changed the position during motion of a swinging pendulum bob in a coupled oscillator experiment he remembered seeing; he inquires. An interesting one from an upstate New York high school teacher asks about batteries for his Ruhmkorff coil, his model producing a healthy six inch spark (clearly not Queen & Co.'s finest). He further asks: "Is there any danger of an explosion in running the wireless telegraph apparatus in the same room with gas and common chemicals, including gun powder?" From Tubingen, Dr. S. H. Bucherer (one of the early investigators to find e/m to depend on velocity) writes concerning an article for which he had been asked. He fears it is a little too chemical, so it apparently never was done but he goes on: "... glad to hear matters are progressing so satisfactorily at Cornell. Does the University not suffer from the bad times in the U.S.? Germany is exceedingly prosperous and is getting wealthier from year to year. This is due to good government." Child sends some interesting and long letters, handwritten from abroad; he has seen J. J. Thompson, who has helped him during a stay of a month or so at the Cavendish Laboratory. He describes the nice Zeeman experiment shown him by Elster and Geitel, that of viewing two sodium flames, one through the other, one of which as a filter is between the poles of an electromagnet. The effect of the magnet is not great but clearly observable.

There is this rather interesting exchange of letters: From City College in New York, F. G. Reynolds writes Merritt concerning his "Viscosity of Air" paper (the topic will come to run through this "history") to which a Professor Saurel has referred in a note to the editor. Later, Saurel, in reply to a query from Merritt related to the submission, allows as how he "never read the thesis—the atomic theory of gases is to me a sealed book." Saurel has it from Reynolds that the formulae employed are "quite orthodox, taken without change from the memoirs of Maxwell and Meyer." The paper was obviously Reynolds' thesis, on which he got little help and "was allowed to work out his own salvation." Which he did, and the paper was accepted. This would make a nice story: could he be *the* Reynolds, as was suggested, he of the dimensionless number in viscous fluid flow? Reference to a science biographical encyclopedia dampens the enthusiasm; that Reynolds, Osborne, was British, not surprisingly, reasonably contemporary with our man, F. G., however.

Late in 1910, Bedell wrote a letter to botanist Professor F. C. Newcombe at Ann Arbor about the REVIEW; Newcombe was contemplating a journal. For what it says about the REVIEW operations, we give the letter in full:

Dear Professor Newcombe:

With reference to your inquiry concerning the finances of the PHYSICAL REVIEW there has always been a deficit which has been borne by Cornell University. The REVIEW was established in 1893 and for a number of years the annual deficit was in the neighborhood of $2000, sometimes more and sometimes less. The annual deficit has gradually been reduced by increased circulation and by more economical production. At the start the REVIEW was published by Macmillan Company for Cornell University, the Macmillan Company being merely agents. The printing was first done by most expensive printers and all numbers were cast. Later we printed from type and employed less but perfectly satisfactory printers.

The deficit of late years has been less than $500 per year and some years we have made no call at all upon the University. This, however, has been a factitious (sic) showing, for large bills for *illustrations, paper, etc. sometimes come in irregularly and may by chance come in just after the close of a fiscal year.*

Since last January we have been independent of the Macmillan Company and conduct all business from this office. This will result in economy in as much as we save the commission of I think 15 percent hitherto paid the Macmillan Company and it will likewise result in simplicity for we now deal directly with engravers, printers, etc. which saves some confusion and loss of motion.

A number of years ago the circulation of the REVIEW was considerably increased by an arrangement made with the Macmillan Physical Society (sic) whereby all members received the REVIEW, the Society paying the $3.00 per year to the REVIEW. The subscription to the REVIEW was at that time $5.00 per year and has since been increased to $6.00. (Does he not mean the Society dues?)

With the increased circulation due to this and other causes the REVIEW would more quickly reach a position of independence were it not for the fact of the increased bulk of material requiring publication. This increase is now so rapid that although gratifying in some respects it creates a serious financial problem.

We have always printed more than enough for immediate circulation and the sale of back numbers and occasional complete sets brings us in sometimes several hundred dollars in the year.

In as much as it has not been a year since our taking over the management from the Macmillan Company it is not easy to state exactly our present condition but it is substantially as follows:

If we issue about 1300 pages per year and print about 1300 copies of each issue our expenses will be about $4000 per year or a little over. (Our expenses for the first six months of 1910 amounted to about $2300.) Our income from 300 subscribers and 600 members of the American Physical Society will be about $3400. Miscellaneous receipts of extra reprints, odd numbers, etc. give an irregular return of several hundred dollars.

Most of our subscriptions come through agents who receive a commission of one sixth.

We are very glad to place this information in your hands for you to use discreetly in considering the question of botanical publications. You will consider of course that it is to a certain extent personal. We will be glad indeed if it may be of an assistance to you.

Yours, very truly,

Bedell (a scrawl)

The copy was probably not actually that sent to Michigan, for some errors clearly remain.

Germane to the letter is a sheet found in the records giving the following financial statements:

PHYSICAL REVIEW
Vol. I
Aug. 1, 1893 (and before) to Aug. 1, 1894

Annual appropriation	$2400.00
Credit: Macmillan & Co.	405.65
Credit: Nalder's advt.	125.00
	2930.65
Expenditures	2056.19
Unexpended Balance Aug. 1, 1894	$874.46

Vol. II
Aug. 1, 1894 to Aug. 1, 1895

Reappropriation of unexpended balance	$874.46
Annual appropriation	1500.00
Special appropriation	546.95
Credit: London sales	63.05
Credit Macmillan July 1895	384.82
	3369.28
Expenditures	2985.41
Unexpended balance Aug. 1, 1895	$383.87

Vol. III
Aug. 1, 1895 to Aug. 1, 1896

Reappropriation of unexpended balance	$383.87
Annual appropriation	$2016.13

And that's it!

In the Volume I accounting there is no note made of the $500 appropriation we have cited earlier; it is perhaps covered in the $2400 amount listed. In Volume II there is listed $546.95 as a special appropriation. One finds in the records a letter from Frederick Bedell to the Treasurer (June 15, 1895) a request for that amount "for payment of bills now due." And a memo from the Treasurer appropriating the amount, signed again in great style by E. L. Williams.

Why the statement on Volume III was not completed is anyone's guess. The Volume I reference to "Credit: Nalders advt." is puzzling. The first Nalder's ad found in the Cornell collection of Reviews comes in Volume II, 1894-1895, when instrument advertising seems first to appear. There is no such credit noted for the following year.

Finally, in this casual survey of early Physical Review office records, one might mention one rather nice piece of technology used in the office, awkward in execution but in neatness and in ease of reference in some ways superior to what is around today. There is with the Review documents at Cornell, a large hard bound volume containing 1000 letter-sized sheets of thin tissue, on each of which in blue type (or black, if handwritten) a copy of a letter is preserved. Each page is numbered at the upper right corner in big blue numerals, consecutively. Provision is made for easy indexing with indentions on the sides of hard paper pages up front. Look in the index under "W" for Williams; page numbers of correspondence to this character are given, and there he is on page so and so appropriating $546.95. Letters are not lost unless the whole thing disappears. And it is easy to reference. In spite of directions on the inside of the front cover, how it was done remained unclear until explained by the University Archivist. There was provided two oiled sheets. The secretary places one behind the tissue on which a letter is to be copied, the tissue is moistened with water and closed down on the oiled sheet with the letter to be copied between, face up with the typed material next to the moist tissue. Another oiled sheet is then placed atop this, and the whole sandwich is squeezed by placing the book in a press. The typed script is water soluble enough that some transfers in the process, coming through the thin tissue to be read by posterity. It is a convenient, long gone technique for storing one's correspondence for future easy retrieval. Today, floppy discs, 64 K memories, and cathode ray displays do it faster and perhaps better; so the writer hears.

Chapter 5
A NEW SOCIETY

·◆·

An event of the period took place in 1899 which was most significant to the PHYSICAL REVIEW and to the world of physics generally. This was the formation of the American Physical Society. The impetus for the move came from A. G. Webster of Clark University at Worcester. He was somewhat remote up there and strongly felt the need for a forum devoted solely to physics. Section B of the A.A.A.S. meetings was not quite adequate. Nor were the sessions of the National Academy, to which Webster did not then belong; he wasn't sure he'd find them enjoyable even if he did. He was not alone in his feelings. He obviously discussed the notion with a small group and sent word out more generally setting up a meeting at Columbia University further to look into the idea. The letter follows:

Dear Sir:

The American Mathematical Society has been in existence for over ten years, and has been markedly successful in bringing together mathematicians and influencing the growth of mathematics in this country. Many other professional societies might be mentioned with similar aims. It seems to us that the time has now arrived for the organization of physicists into a national society, which shall be for this country what the Physical Society is for England, and the Deutsche Physikalische Gesellschaft for Germany. We propose a society meeting four or more times yearly in New York for the reading and discussion of papers. Of the advantages of such a society there is small need to speak. Few things inure more to the advantage of scientists than frequent meetings for the purpose of interchange of ideas and learning of one another's work. Such meetings have so far been too infrequent. Persons who have attended meetings of the American Association for the Advancement of Science must have wished that such opportunities for social and scientific intercourse were more frequent, as well as possible at other times of the year than in the

77

summer. An organization like the one proposed could not fail to have an important influence in all matters affecting the interests of physicists, whether in connection with work done under Government auspices or otherwise.

The proposed society will conflict with the interests of no other organization, and will represent no institution or clique, but will be devoted to the advancement of our science. We invite you to attend a meeting to be held at Columbia University, New York City, on Saturday, May 20, at 10 a.m., in Fayerweather Hall, for the purpose of discussing this matter, and if possible, organizing a Physical Society. If you are unable to attend, will you not communicate with one of the undersigned, making any suggestions that you may think important, and signifying your desire to join and support such a society.

> A. G. Webster, Worcester, MA
> J. S. Ames, Baltimore, MD
> E. L. Nichols, Ithaca, NY
> C. Barus, Providence, RI
> M. I. Pupin, New York, NY
> B. O. Pierce, Cambridge, MA
> W. F. Magie, Princeton, NJ

The aforementioned chemists' expedition to the Cornell Physics' attic turned up cards to be mailed to various institutions announcing again the meeting:

Meet in Room 304, Fayerweather Hall of Columbia University at 10 a.m., May 20, 1899, to consider a possible physical society.

Endorsed by: the same list as above, with institutions.

Melba Phillips in *Physics Today* for June 1987 has recounted something of Webster's career and his role in founding the Society. Merritt, writing in the *Review of Scientific Instruments* of April 1934, and Bedell in the PHYSICAL REVIEW on its fiftieth anniversary (Volume 75, 1601, 1949) and in an invited paper at the Berkeley APS meeting of 1949, both give their versions of the founding. Merritt notes the role of Section B of the A.A.A.S.:

Previous to the establishment of the Physical Society, Section B of the American Association for the Advancement of Science was practically the only place where physicists from all over the country could get together for discussion and for presentation of papers. Most of the physicists of thirty five years ago recognized their indebtedness to the Section for the inspiration they received at its meetings and for opportunities that it gave for becoming acquainted with others.

There was a very friendly feeling toward the Association, and no desire to compete; rather it was to supplement the work of Section B. A resolution adopted at the first APS meeting called for cultivation of closest relations with the Section and "to contribute everything in its power to the latter's success." Merritt discusses the PHYSICAL REVIEW and its role in the Physical Society:

In the years 1893 to 1899 the PHYSICAL REVIEW undoubtedly contributed in no small degree to the increased activity (in physics) which later resulted in the establishment of the Physical Society.

Bedell describes the Society founding:

After a welcome from Professor M. I. Pupin, and a discussion of aims and policies, it was resolved that a physical society be organized. A committee was appointed to draft a constitution, officers were elected, and the American Physical Society was born. It was as simple as that.

He stresses the importance of the 1876 Centennial Exposition in promotion of a scientific climate in America. This was the same year that Henry Rowland was appointed Professor of Physics at the new Johns Hopkins. His work was recognized by Maxwell and other European physicists and he was seen as the outstanding American physicist, eminently qualified for the Hopkins post. He became a leader in the development of American physics and graduate research. Thus, twenty years later it was entirely fitting that he be elected first President of the new Society. The story is told of Rowland's having to testify in some trial or other and on being questioned as to the greatest American physicist, allowed as how he was. "Isn't that a bit presumptuous," he was asked? "Sir, I am on oath," was the reply. Something like that. (The tale may be apocryphal; see two letters in *Physics Today* of 1984—Feb., p. 89 and Apr., p. 109.) Looking through his collected papers, one may conclude that he was not far off the mark.

In his reminiscence, Bedell wrote: "The need in this country for special societies in various fields was becoming more and more obvious." It was not clear what connection there might be between such special groups and the A.A.A.S., where the matter was frequently discussed. According to Bedell, the same questions were posed in England, vis-a-vis the British Association for the Advancement of Science. There the growth of science in the United States was recognized and serious consideration was given to the Physical Society of London's setting up an American section. The need for some sort of American physics society was clear on both sides of the Atlantic. Webster took the initiative and as chairman of a "representative committee," which he formed, sent out the call for the Saturday, New York, meeting with the results cited. The "representative committee" consisted of the men signing on the above prospectus and follow-up card mailed to likely members.

Nearly forty physicists attended the meeting in Room 304, Fayerweather Hall. Included were the leading lights: Rowland (in spirit but actually in absentia) and Michelson, in company of such as Webster, Nichols, Merritt, Bedell, Magie, and Abbe (alone from government). Gibbs at Yale never did sign on. A letter from Magie in the Cornell Merritt file tells of some of what went on. Magie practically wrote the constitution himself, modeling it on that of the American Mathematical Society. He recalled Webster's promoting the first officers: Rowland, president, Michelson, vice-president, Merritt, Secretary, Wm. A. Hallock, treasurer, and various others on the Council, including Cornell's E. L. Nichols. Merritt, as we have noted, would in one capacity or another, Secretary in the formative years arranging programs and issuing the quarterly *Bulletins*, and later as President and Councilor, serve the Society for almost fifty years. Withal this service, he was also concurrently for thirteen years an editor and frequent contributor to the PHYSICAL REVIEW, which during that time continued to publish under the auspices of Cornell. But the future was promising for the journal; more was being published, a Society was almost alongside (definitely not yet associated) and the number of physicists was growing.

The second meeting of the Physical Society followed in due course that of the organizing session, again in Fayerweather 304, Columbia, in October of the same year. Rowland, ill with diabetes, gave a stirring, and oft referred to but obscure, presidential speech, "The Highest Aim of the Physicist," something of a swan song for him; he died two years later. He seems rather bitter but is encouraged over the future. He again pleads for "pure" science:

> ... *Thus we meet together for mutual sympathy and the interchange of knowledge, and may we do so ever with appreciation of the benefits to ourselves and possibly to our science. Above all, let us cultivate the idea of the dignity of our pursuit so that this feeling may sustain us in the midst of a world which gives its highest praise, not to the investigation in the pure ethereal physics which our Society is formed to cultivate, but to the*

one who uses it for satisfying the physical rather than the intellectual needs of mankind. He who makes two blades of grass grow where one grew before is the benefactor of mankind; but he who obscurely worked to find the laws of such growth is the intellectual superior as well as the greater benefactor to mankind.

How stands our country, then in this respect? My answer must still be now, as it was fifteen years ago, that much of the intellect of the country is still wasted in the pursuit of so-called practical science which ministers to our physical needs but little thought and money is given to the grander portion of the subject which appeals to our intellect alone. But your presence here gives evidence that such a condition is not to last forever.

Why all this puffery about superior intellect? Is it not enough that we are just trying to find out how the universe works? He winds up:

But the true lover of physics needs no such spur (the conquest of disease he has cited) to his actions. The cure of disease is a very important object and nothing can be nobler than a life devoted to its cure.

The aims of the physicist, however, are in part purely intellectual; he strives to understand the universe on account of the intellectual pleasure derived from the pursuit, but he is upheld in it by the knowledge that the study of nature's secrets is the ordained method by which the greatest good and happiness shall finally come to the human race.

So we do need our Edison's, our Morse's, and our Bell's.

Where then are the great laboratories of research in this city, in this country, nay, the world? We see a few miserable structures here and there occupied by a few starving professors who are nobly striving to do the best with the feeble means at their disposal. But where in the world is the institute of pure research in any department of science with an income of $100,000,000 per year? Where can the discoverer in pure science earn more than the wages of a day laborer or cook? But $100,000,000 per year is but the price of an army or a navy designed to kill other people. Just think of it, that one percent of this sum seems to most people too great to save our children and descendents from misery and even death!

But the twentieth century is near—may we not hope for better things before its end? May we not hope to influence the public in this direction?

May we, indeed? It is perhaps as well that he could not know what that century held for mankind.

> *Let us go forward, then, with confidence in the dignity of our pursuit. Let us hold our heads high with a pure conscience while we seek the truth, and may the American Physical Society do its share now and in generations to come in trying to unravel the great problems of the constitution and laws of the universe.*

And that we seem pretty well to have been doing the past eighty or ninety years.

At this second meeting of the Society, Magie's constitution was adopted and there were three papers given—by Rowland in his presidential address, by Pupin, and by Webster. By year's end there were all of 59 members in the Society; Room 304 Fayerweather was adequate to hold the meeting participants for years. In fact, it was something of a problem for Merritt, as secretary, to organize a program, on occasion speakers on the program only learning of their assignment in a telegram from Ithaca, some indeed refusing last minute assignment, their papers given in title only. It was of some embarrassment to Merritt that once in a while he had to throw in something of his own to fill a void.

The constitution consisted of six Articles and a dozen By-Laws. The second Article states the aim of the Society (dubbed the American Physical Society in Article I) as: "The object of the Society shall be to promote the advancement and diffusion of the knowledge of physics." In these days of a very complex world, there have surely been some amendments (Article VI) to allow of other activities, more sociological or political, now carried on by the Society (and questioned by many). By-Law II decrees the annual dues to be five dollars "payable on the first of January," and penalty of nonpayment ("Removal of his name from the list of members after due notice." It is not decreed what happens if per chance his name should be her name.) For fifty dollars one could become a life member (Article III). There has been some extensive mending on this score over the years, the first not too much later when dues were raised to six dollars. Four meetings a year are to be held, and various other procedures, including the provision for a Council, are spelled out in the document.

For years after 1909 the APS met jointly with Section B of the A.A.A.S. in its annual meeting. The first APS meeting outside of Fayerweather 304 was in 1904 in Washington; twenty papers given. In 1905 they tried with success one in Chicago. For a long time after that there were yearly meetings in both places. Local sections came later but the desirability of local groups was recognized in a resolution passed at the first meeting.

The PHYSICAL REVIEW did not immediately note the formation of the Society. The Society had its own publication—the *Bulletin of the American Physical Society*, edited by Ames, Pupin, and Merritt; issued quarterly. It contained the abstracts of papers presented at meetings and minutes of the Society business sessions. Circulation was almost limited to Society members—perhaps 150 copies printed and mailed. But in its brief life it carried some papers of note: Michelson on the velocity of light, Pupin's loading coil theory, Lyman on spectrum ghosts, Nichols and Hull on light pressure, and Rowland's presidential address. Merritt wrote later of knowing of but four complete sets of the *Bulletin*, covering APS meetings Number 1 through Number 7. None of the four sets is at Cornell, where at least one of them might have been expected; holdings there include only Volumes I and II (donated by Merritt) of the three that were published. Printing was done by the Ithaca press of Andrus and Church, established and venerable when Cornell opened its doors, now extinct.

The *Bulletin* lasted but three years. APS paper abstracts were thereafter published in the REVIEW; Society members could subscribe to the journal at a discount. (With no Volume III of the *Bulletin* at Cornell, and not even knowing where one is, there is no record of the APS minutes for the business meeting at which these arrangements were set. Publication of the *Bulletin* was apparently resumed in 1925. The Cornell Library holdings only include Volumes 1–, 1956–, of Series II. The catalogue card there indicates the *Bulletin* dates from Feb. 14, 1925; in a Series I, Volumes 1 through 30 were published, with a footnote on p. 4 (of Volume 1 presumably): "The present *Bulletin* is not a continuation of the *Bulletin of the American Physical Society* which was first published in 1899 but discontinued after a few numbers." No Minutes of APS Meetings Proceedings seem to make note of the resumption of publication.) Volume XVI of the PHYSICAL REVIEW, however, observes this change and carries the front page note that the journal is conducted with the cooperation of the American Physical Society by our three editors, no longer indicating that it was being published for Cornell University, although it would be carried by the University for another ten years. Macmillan would disappear with Volume XXXI in 1910, in favor of the PHYSICAL REVIEW, Lancaster, PA, and Ithaca, NY. That there was indication of a European representative (Berlin: Meyer and Mueller) noted with the first issue, was dropped in Volume II of Series II in 1913.

The first recognition by the REVIEW that there was a new Society of physics seems to have come in the year after the founding, in

Volume XI, in a summary of the summer A.A.A.S. meeting in New York City. Included in the summary, probably written by Merritt, was this paragraph:

> *For Section B the meeting was especially significant on account of the cooperation of the American Physical Society. The sessions of Section B and of the Physical Society were held on alternate days. But although the programs were kept separate, there was little else to distinguish the two. The holding of such a joint meeting, for the first time in the history of Section B, was very properly regarded as an experiment. It is hoped that its undoubted success will lead to the permanent enrollment of the Physical Society among the numerous organizations already affiliated with the Association. The great increase in the scientific activity of our American physicists has now made the existence of the two societies not only possible but desirable. Certainly the cooperation of the two societies can scarcely fail to be of mutual benefit.*

The summary listed the program of papers of Section B and the fifteen in the Physical Society sessions, several contributors appearing on both programs, some multiply so.

Chapter 6
THE NEXT DECADE

••

For the PHYSICAL REVIEW the decade 1900 to 1910 or so appears rather uneventful. The journal was being subsidized throughout the period by Cornell but by the end of the decade, through cutting costs to the bone, it was showing small profit. There was still an inordinate number of papers by Cornellians but the scope of authorship was broadening, the volumes were getting fatter (with no change of paper weight) and, with Volume XVII, the pages cut and trimmed.

Volume XVI (1903) began printing the Proceedings of the American Physical Society meetings, with Minutes where business was conducted but almost exclusively abstracts (or Title Only) of the presented papers, the first printed Proceedings those of the Eighteenth meeting, held in Washington at the Columbian (?) University. A footnote marks the change: "The minutes of all meetings of the American Physical Society will hereafter be published in the PHYSICAL REVIEW. The publication of the *Bulletin of the American Physical Society* has been discontinued." Abstracts did not always appear in the same issue carrying the Proceedings and Program, a somewhat disconcerting situation. The flyleaf for the volume carries notice that the journal is "Conducted with the Cooperation of the American Physical Society" by our three editors. No more mention that it is published for Cornell; still by Macmillan, however.

It is always fascinating to peruse journals of an earlier time. One runs into people doing work for which they later become famous, or work quite different from that which gave them their fame. There are amusing (to us) titles of papers, landmark papers, papers that are harbingers of things to come, notable events in physics, locally familiar names not generally so well known, and so on. This is certainly the case for the present writer with the early PHYSICAL REVIEW. Not only are the paper titles (and, more, some papers) of interest but perhaps even more so are the titles and abstracts of papers given at the Physical Society meetings which appear in the Proceedings carried by the journal. There are many more of the latter to be sure, so that one is more likely there to strike gold. Moreover, if a researcher is on to something good, he/she will publish the full blown paper where it will reach the largest and most widely recognized audience, which in these early years was certainly not the list of subscribers to the PHYSICAL REVIEW. Thus, considerable gold went to European publications.

While the struggling journal may have been somewhat in the doldrums during the first decade of the century, it still seems appropriate to record a few of the more interesting items the REVIEW noted and published, recognizing that what appears interesting to one person is not necessarily that which might interest someone else. For whatever it is worth, we nevertheless record what caught the writer's eye in his looking through back issues of our journal.

And the year 1900 seems a good year to begin our perusal; it opens the twentieth century and the year marks the dawn of quantum theory, with Planck's introducing the photon for the explanation of black body radiation. Abstract Number 507 of *Science Abstracts* for 1901 is "Distribution of Energy in the Normal Spectrum," M. Planck, published the previous year in *Deutschen Physikaleschen Gesellschaft*. In that year, the REVIEW published its Volumes X and XI, the second of which carried the previously mentioned note concerning the first joint meeting of the American Physical Society and Section B of the A.A.A.S. It also included a memoir and frontispiece portrait marking the death of Thomas Preston, best known for his book on light. O. M. Stewart, at Cornell, had another long resume on the literature pertaining to Becquerel rays (that research was growing); from Stanford, F. J. Rogers (not William A.) writes on the advantages, in teaching, of a system of units based on the meter, the kilogram, and the second, "or briefly, the m.k.s. system," in his words. In another paper, Shedd, at Wisconsin, discusses the appearance in shape of the fringes in Michelson's interferometer which Michelson had introduced eight years earlier in the *Philosophical Magazine*. That was the kind of gold which, for meaningful exposure, generally went across the ocean for publication.

In the preceding volume, X, there was a paper by C. D. Child on "A Dissociation Theory of the Electric Arc," one on the "Influence of Electrification on the Surface Tension of Water and Mercury" by

Merritt and S. J. Barnett (we'll be hearing from him again, more importantly), and one by Bevier at Rutgers on the analysis of vowel sounds, using the phonographic wax cylinder of Edison's commercial reproducer, in development twenty years previously "much less true to timbre than it is today," as Bevier notes. He mounts a small mirror on the stylus to reflect a light beam to the recording chart, magnifying the groove undulations some 1000 times. R. A. Fessenden writes on the density and elasticity of the ether, his second contribution concerning the medium; in Volume I he wrote on its dynamics. He would be better known for his developments in radio telephony; in 1906, on Christmas Eve, he would make the first American radio broadcast, from his experimental station in Massachusetts, the program consisting of a poetic reading and short talk. Before his death in 1932 he held over 300 radio related patents.

In 1901, Volume XII of the journal carries another frontispiece portrait, this time of G. F. Fitzgerald, who has died at the age of a mere 50 years; a long paper by Larmor discusses the man's contributions, mentions of course "Michelson's classical interference experiment," and gives credit to Fitzgerald as the first to suggest that cohesive forces might be affected by motion. Curiously, there is also a book review of the Lorentz Jubilee Volume, where the great memoir on his theory of electrical and optical phenomena in moving bodies is stressed: "... the development of the theory at the hands of writers (named) leads to a number of startling conclusions... ," among them that gravitational disturbances are transmitted at light velocity and that "mass as defined in terms of inertia is constant only for small velocities and increases with the velocity of the moving body." Einstein was still four years in the offing. Contributed papers in Volume XII include another by Child on the "Velocity of Ions Drawn from the Electric Arc," in a method suggested by J. J. Thompson; one by Lyman on "False Spectra from the Rowland Concave Grating"; and by Austin, one on "The Application of the Manometric Flame to the Telephone." Fortunately for us today this last was not the development of a needed accessory critical to telephone use. Rather it was a study using the flame to learn what a receiver diaphragm was doing. Viewed in a rotating mirror, the flame height variations made a primitive oscilloscope; Merritt's paper in Volume I and Bevier's later made use of it in their "vowel" studies.

The death of the dean of American physics, Henry Rowland, was noted in Volume XIII (1901)—another frontispiece and a long memoir by the REVIEW Editor-in-chief. In it he discusses at length the controversial and tough experiment of Rowland's on the magnetic effect of a revolving charge distribution, on which Nichols had also worked at Johns Hopkins, as we have remarked. He cites Rowland's pleas for pure science espoused so strongly in the A.A.A.S. speech, and reiterated in the first APS presidential address. Visible radiation of glowing carbon is of concern at Cornell and elsewhere; Nichols has two papers as sole author in this volume, another with Blaker, and Blaker has a *Note* by himself—all pertaining to this source of radiation. And in the same vein, G. W. Stewart (brother to O. M.) is looking at the energy distribution in the spectrum of an acetylene flame. P. G. Nutting writes on the "The Metallic Reflection of Ultra Violet Radiation." This matter would be of more than passing interest in thirty years time by other Cornell (and Johns Hopkins) physicists evaporating aluminum for telescope mirrors. The notable paper in the REVIEW for the year is that of Nichols (E. F.) and Hull on "A Preliminary Communication on the Pressure of Heat and Light Radiation." In the next issue a paper comes by E. Rutherford which is similar to and cites that of Child earlier (on ions from an arc) on the "Discharge of Electricity from Glowing Platinum and the Velocity of the Ions," somewhat off the subject for which he is noted.

The study of vowels for some reason continues to be of interest to Prof. Bevier. In Volume XIV (1902) he has another paper making use again of the track in a phonograph cylinder to look at the "Vowel A^E (as in hat)" and "The Vowel E," analysis by his little wobbling mirror-light beam arrangement. Child is into ions from hot sources with two papers on the "Velocity of Ions from Hot Platinum," following the work of Rutherford above. A new Cavendish balance is developed by G. K. Burgess who discusses measurements of the gravitational constant in a follow-up paper. And then there is one more by E. L. Nichols, "On Some Optical Properties of Asphalt"(!). It turns out that as a film it appears red in transmitted light, cutting off very sharply.

In XV (1903—the year of the Wright brothers) Bevier gives us still two more vowels—"I^E (as in pit)" and "I (as in pique)." S. J. Barnett appears again with a couple of notes on basic electrostatics (Gauss' law and the inverse square law of force in the Cavendish experiment, which his British referee "did not understand"). We have a paper on

"Wireless Telegraphy" by A. H. Taylor at Lansing, Michigan, but one day to be at a future Naval Research Laboratory and figuring in radar development, well before microwaves. He would write frequently over the years from a position in North Dakota, more or less on the same subject. He was in this paper receiving signals at 1 mHz with his simple telephone relay detector, over a four-mile distance between thirty foot vertical antennae. (Except for the simplicity of his receiver not too noteworthy; Marconi had already spanned the Atlantic the previous year.) There was a book review on geometrical optics, the review written by H. C. Lomb; there is in the Cornell Archives a letter to Merritt from Henry Lomb (of Rochester's Bausch and Lomb) declining to review the book: he'd had no college education and the review would be poor. But his son could do it. Okay? Apparently it was. Perhaps the ultimate in something or other was the paper by Frank Allen and William Andler on the "Test of the Liquid Air Plant at Cornell University," done under the auspices of Nichols; the editor must have been hard pressed with that issue. The plant was undoubtedly that which served Ithaca well making oxygen during that typhoid epidemic described by Merritt in our "profile" earlier.

Probably the most significant feature of Volume XVI (1903) is not in the papers published but in the Volume flyleaf. The Macmillan Company still publishes the REVIEW but no longer is it indicated that it is "Published for Cornell University;" it is now "With the Co-operation of the American Physical Society." Strangely, beyond Macmillan's response to notification it received of the proposed tie, nothing concerning this new association has turned up in Cornell records on the REVIEW; and the Minutes of the Society business meetings are silent. (Minutes of the 15th, 16th and 17th meetings have not been found.) The REVIEW's financial situation was improving but Cornell still had losses to cover. By taking this step, the journal picked up some 300 "captive" subscriptions; members of the Physical Society would receive the REVIEW as part of their yearly dues, along with *Science Abstracts*. Beyond this and providing an easily accessible outlet in which to publish work, it is not clear what "Co-operation" entailed. Our journal began publishing the Proceedings and abstracts of the Society meetings, allowing the *Bulletin* to disappear after its Volume III. Both steps served to bring the two enterprises closer together.

Macmillan was encouraging about the plan; in their letter to the editors they are heartily in favor of the proposed change—welcome news in regard to the PHYSICAL REVIEW—the "step a long one and best to be thought of from the standpoint of the REVIEW itself." With greatest pleasure, Macmillan will give up its commission on sales of the REVIEW to members of the American Physical Society (but presumably not on sales to non-members); "... with best good wishes for the future of this very advantageous (from the REVIEW standpoint, etc.) ... and necessary arrangement. Sincerely, ..."

Titles in Volume XVI are more interesting in the Physical Society meeting abstracts than in the regular journal articles, as will generally seem to be the case for quite a while. An abstract of interest is that of Nichols and Hull on their radiation pressure measurements ("final results"), which precedes a perhaps even more important one by E. Rutherford on the "Magnetic and Electric Deflection of the Easily Absorbed Rays from Radium." He is in his decade at McGill and on to a good thing, finding his alpha rays to be "positively charged bodies projected at a velocity of about one tenth that of light and of mass of the same order of the hydrogen atom and large compared to the electron." There are other "radioactive" abstracts; one by Rutherford and Cooke on a radiation of great penetrating power, "appears to come equally from all directions," getting through 5 cm of lead, indeed even when "5 tons of pig lead was placed around the apparatus." There was another by McLennen on the radioactivity at waterfalls (Niagara) and a similar one by Allen looking at the radioactivity of snow. It was all a puzzling business. Merritt ends his Minutes of the 20th APS meeting (Fayerweather 304) with this paragraph: "At the close of the afternoon session, Mr. H. T. Barnes showed to the Society an experiment first described by Sir William Crookes (a year earlier in 1903 to the Royal Society) in which microscopic examination shows the luminescence of zinc sulphide produced by radium is irregularly distributed over the field in a number of starlike bright spots, each of which lasts but a short time. It had been suggested by Sir William Crookes that each luminous spot was due to the impact of a single corpuscle emitted by the radium." Geiger and Marsden would become very familiar with scintillations. The first paper of the volume was perhaps the most interesting contribution, on a rather obscure effect; Roller R. Ramsey (R^3) of Cornell, working under Nichols, studies the effect of gravity (and pressure) on electrolytic action. It was suggested (and looked for unsuccessfully) by Maxwell that a counter EMF should exist in electrolysis depending on whether the plated ions move up or down in the gravitational field.

Ramsey observes roughly the appropriate number of microvolts. A book reviewed by Bedell was that we noted in Macmillan's REVIEW advertisement, of Jackson and Jackson, "An Elementary Book on Electricity and Magnetism and their Applications." "Inspiring," the review notes. Neither Jackson was author of a much later text on the subject, one widely used by graduate students at the end of the century, and not so elementary. Dying during this year was Ogden Nicholas Rood (yet another Nichols, almost); the biographical note was again by E.L.N., our man. It is probable that he did not know of Rood's opinion of Cornell's founder, cited earlier. The frontispiece portrait of Rood rather fits the picture.

The following REVIEW volume, XVII, in 1904, carries notice of a greater loss to science. There is the long memoir by Joseph Trevor, our shy co-editor of the *Journal of Physical Chemistry*, on J. Willard Gibbs, at his death still rather the loner there at Yale. Gibbs never did publish anything in the REVIEW. For some reason, inappropriately, there was no sepia portrait this time. A new book is reviewed: "Light Waves and their Uses" by A. A. Michelson, the book well known even today by students in optics. In two papers of a trilogy, Harvard's E. H. Hall, discoverer in 1879 (!) of the effect bearing his name, discusses the matter "Do Falling Bodies Move South?," citing early controversy (as far back as Newton) and his recent experiments in the tower at Jefferson Physical Laboratory at Harvard. His "positive" result is hardly out of the noise. In his first paper he feels more investigation should be carried out "so that the question before us may vex no mortal more," agreeing with Cajori that the ideal laboratory was to "be found" in the "great pile" in the "Great monument at Washington." (More significant to gravitational matters and less subject to hydrodynamic effects and shaky initial conditions would be the experiment of Rebka and Pound in the much more recent "Mossbauer era," showing the gravitational red shift of Einstein in an experiment conducted in the same Jefferson tower.) Also in Volume XVII the important work of Nichols and Hull is reported in full detail in two papers in the successive issues of July and August. Between the two is a paper by Coblentz at Cornell on the optical properties of iodine, and one of work started under Nernst, by Hartman, now gone to Pennsylvania, on the visible spectrophotometry of the Nernst glower, a better source than the acetylene-hydrogen flame and one used widely in infrared work even today, but apparently never studied spectrally by Nernst. Nutting, at Cornell, continues his work in the ultra-violet with a paper on "Ultra-violet Rotary Dispersion," but more sensationally, investigates the "Distribution of Motion in a Conducting Gas," studying the heating with a thermopile effectively moved along a discharge. He makes good use of Cornell's Dynamo Laboratory, employing a gang of up to 24 (!) 500 volt DC dynamos connected in series. Eight was "best" and "... more easily controllable." "12,000 volts continuous leaks badly over wood, into air, and across switches. The current (not so small) was controlled by a cadmium iodide in amyl alcohol resistance." High voltage technology; one can imagine the noisy arrangement. G. B. Pegram, long to be at Columbia, is into nuclear physics with "Secondary Radioactivity in the Electrolysis of Thorium Solutions," and Fernando Sanford, "On an Undescribed Form of Radiation" elaborates on that paper he had in Volume II, on "electric photography." Many strange effects; curious.

The contributed papers of 1904 (Volumes XVIII and XIX) are for the most part not very interesting. T. E. Doubt, on suggestion of Michelson, checks the effect of intensity on the velocity of light in air, water and CS_2; "... to use the interferometer in a similar way to that used by Michelson and Morley in their investigation of the motion of the medium upon the velocity of light," namely, the Fresnel drag which they measured three years after their ether experiment. Over an intensity range of 40,000 to one, the velocity is not affected by "one part in 1000 million." And John Mills at Nebraska checks the effect of a magnetic field on the velocity of light in a liquid—the Faraday effect. He uses a Michelson interferometer with, in each arm, a 75 cm tube of CS_2 surrounded by a coil. With a Nicol prism and a Bravais double plate, right-handed light is present in one half the field of view and left-handed light over the other half, both circularly polarized. Connecting the two coils in opposing sense, the fringes moved one way in half the view and the other way in the other half by an amount linear with coil current. Rather pretty. In Volume VII, Eddy, Morley, and Miller's similar experiment with plane polarized light had used the identical instrument and found, as we noted, no change in velocity with the field. (It will be recalled that Miller, skeptical of the negative result of the ether drift experiment, would repeat that experiment and would long hold that there was a positive fringe shift.) D. C. Brace was instigator of Mill's work and followed the paper with one of his own on the "Half Shade Elliptical Polarizer and Compensator," with which he would check the Lorentz-Fitzgerald contraction hypothesis, with the expected, today's result. In Volume XX, Eddy (Minnesota) has an electromagnetic theory to cover the results of the Mills and Brace experiment. Somehow, the editors found liquid air production of interest; we have (Volume XIX) another test of a liquid air plant, this time at Wesleyan University. A couple of the APS meeting abstracts in XVIII were noteworthy: Theodore Lyman of Harvard reported on measurements of the short wavelength radiation discovered by Schumann, and Rutherford had one on "The Heating Effect of the Radium Emanation." In Volume XX, in the next year, Lyman would have another and short one with special emphasis on hydrogen, in which, encroaching on the as yet unrecognized primary series of the first element, his range of wavelengths measured covers the first two members of the series.

More interesting in the Proceedings during 1904 in Volume XVIII is the presidential address of A. G. Webster, "Father" of the Physical Society, to the Society in a meeting at the scene of the founding. "Some Practical Aspects on the Relationship Between Physics and Mathematics" is rather nice, even the twenty pages of it. He cites the loss of Rowland and Gibbs during the first four years of the Society, giving particular attention at length to Gibbs'contributions. He cites gains: Nichols and Hull's radiation pressure work, the confirmation before his death of Rowland's famous and controversial electric-magnetic effect, the achievement of flight and the creation (1901) of the Bureau of Standards under Stratton, who had vigorously pushed it. (The Society had already, at its fourth meeting, appointed a committee to "draw up a Memorial to Congress on behalf of the Physical Society urging the establishment of a Bureau of Weights and Measures"and, in transmitting the Memorial, expressed its "cordial approval of the bill pending in Congress for the establishment of a Bureau." See Merritt on the founding of the APS, referred to earlier.)

Like Rowland before him, Webster bemoans the small contributions made to physics by Americans, and ventures the reasons: the taming of the country, the seducements offered by commercial success, material prosperity, the loss of ideality, and importantly, the lack of mathematics in physics training, considering this at length. He sketches out a curriculum, which includes considerable "culture." What little he knows of Greek and Latin, he "would not exchange for any number of definite integrals, or even milligrams of radium." We "need love of the beautiful and of the ideal." Let not the poet berate us for our interests and pursuits. "Is the music less sweet because we can resolve its motion into harmonics and develop it in a series of normal functions?" Both aspects have their place—fact and fancy. He admits that Tennyson's

> *Every moment dies a man*
> *Every moment one is born.*

is not improved by Babbage's then more accurate

> *Every moment dies a man,*
> *Every moment one and a sixteenth is born.*

He concludes hoping that the Physical Society and the Mathematical Society "ever continue their ways like twin sisters, hand in hand, in pursuit of the good, the true, and the beautiful." It is distressing to realize that this man would be a suicide twenty years later.

It is perhaps of interest that in the Minutes of this APS Columbia meeting, announcement was made of the Spring meeting to be held in Washington. Featured would be "Saturday afternoon an excursion to the Bureau of Standards and the Weather Bureau." It would not be until the Spring meeting of 1906, however, that the Spring meetings would all be held at the Bureau, starting a tradition that went well beyond World War II, after which larger facilities became essential.

1905 was a memorable year in physics—particularly for *Annalen der Physik*; in it were published in the short period the amazing papers of Albert Einstein. APS members would learn of his relativity in their *Science Abstracts* (Number 2277): "Electrodynamics of Moving Bodies." For the PHYSICAL REVIEW, the year was not particularly notable; in the year's volumes (XX and XXI) there are papers by Michigan's H. M. Randall, Cornell's Richtmyer (his first of many to come) and Fenner, and a fairly significant one by R. W. Wood on the "Magnetic Rotation of Sodium Vapor." Bevier is still reporting on vowels, Pierce on wireless circuits, and the Cornell crew on a variety of minor subjects. Thos. C. Hebb at Chicago measures the velocity of sound in a method suggested by Prof. Michelson, rather analogous to the latter's velocity of light method. Bergen Davis at Columbia's Phoenix (!?) Laboratories reports on the electrodeless discharge, known by this time. Rankin discusses the use of a magnetic coil around the neck of a Ryan-Braun cathode ray tube, which among other things improves the focus—in reality a magnetic lens. A rather important and well known optics paper first appears as an APS meeting abstract: A. B. Porter's lovely experiments on Abbe's theory of microscopic vision. But then, subsequently and reasonably, he reports fully on it in the British Philosophical Magazine. This was still a frequent occurrence, one our journal editors well recognized and hoped would change. A book is reviewed of lasting importance in its subsequent revisions: Rutherford's Radioactivity. In reviewing a number of books on the subject (by Soddy, by Strutt, by Mme. Curie, and Jacques Danne), Merritt, the reviewer, gives Rutherford high praise, Strutt and Soddy, both favorable notice, and the French, well, so-so.

Citation of Porter above points out a serious defect in some of these early papers: where did the man do his work? (Note added in proof: A chance encounter very recently with an advertisement in REVIEW Volume XXV, Number 5, for the Scientific Shop, sheds a little light on the man. The proprietor is one Albert B. Porter and the shop, on Dearborn Street, Chicago, deals in "Phys. and Astron. Instruments." No clue as to whether he had means for experiment at the establishment or at its works in Evanston.) In the first years of the REVIEW it was frequently the case that articles (and more often, meeting abstracts) were presented with no indication of the author's home base. This changed gradually, but for Porter, there is no clue; even in the *Philosophical Magazine* article, often still referred to, there is no sign. We do better today with the author's institution recorded at the head of the article under his/her name. And an abstract of the article follows immediately. It would not be until 1919 that a "synopsis" of the article was mandated to precede the body of the article itself. A great improvement; and by that time the author's base of operations was universally made known as well.

In Volume XXII (1906) it would appear that the Physical Society has almost taken over the REVIEW, the February issue, for example, consisting of but one paper (Anthony Zeleny at Minnesota), "On the Capacity of Condensers," the rest being devoted to Proceedings of the 30th Physical Society meeting, Fayerweather again, held some time earlier. Granted that most of the Proceedings went to the 28 (!) page presidential address of Brown University's Carl Barus, given at an 11 a.m. session. It must have been a late lunch. More interesting is an abstract of a paper from Rutherford on some properties of alpha rays from radium. Later abstracts of the year include that of Rosa and Dorsey at the Bureau of Standards (already) on their definitive (1 part in 10^4) esu/emu measurement in a determination of c. Of only two papers contributed to the January issue, one is by Coblentz, who has left Cornell to begin his long association with infra-red radiation at the Bureau of Standards, writing on "Infra-red Emission Spectra" and (the next volume) on "Infra-red Absorption and Reflection Spectra," long articles both. Contributions from the Bureau would grow in number and it would begin to play an increasing role in the Society. It was at the Spring meeting in Washington that the Bureau hosted (with the Cosmos Club) an APS meeting, the first in the long series of such Spring Society meetings. In another paper of the volume, L. W. Austin reported on the ejection of negative particles by canal ray bombardment of metal. He had earlier done that piece on gravitational permeability and the one on the telephone and manometric flame. His paper here followed one from Imperial University in Tokyo by Honda and Terada (long before Toyota and Tercel), "On the Geyser at Atomi, Japan."

Things improve in Volume XXIII of the same year. There was the important paper by F. C. Blake and C. R. Fountain from the Phoenix Physical Laboratories of Columbia (when did Phoenix become Pupin?) working on E. F. Nichols' (then at Columbia) seminal suggestion that the selective reflection of resonators and square grids (tiny resonators resulting from a square grid ruled in a metal film) be studied. E. L. Nichols rated the beautiful experiments as second only to those of Hertz in connecting radiation to electricity and magnetism. And a paper by S. J. Allen again, still at Johns Hopkins, on "The Velocity and Ratio e/m for the Primary and Secondary Rays of Radium," confirms Kaufmann's results that e/m depends on velocity when v approaches c. Work by others, however, would not agree. No mention of Einstein, whose relativity had appeared the previous year. Hartman is still with his Nernst glower, trying to get the temperature, and Herbert Ives is improving Wood's diffraction process of color photography, making it into something quite presentable, seen today, however, only in museums. Shedd and Fitch use the interferometer in a cute but straightforward way to obtain the index of refraction of a glass lamina without using a mechanical micrometer on the possibly fragile object. From Berlin S. R. Williams reports on "The Reflection of Cathode Rays from Thin Metallic Films"; the reflection does not all take place at the extreme surface of the metal, but there is a penetration and then reflection for a large part of the reflected rays. A move ahead of Merritt's earlier investigation.

The Physical Society and Section B of the A.A.A.S. met in Ithaca during the year to celebrate the dedication of Cornell's new, then "magnificent" physics building, Rockefeller Hall. Meetings at such special events would be held as well at Princeton (1909), Illinois (1909) and at Yale (1913). Not much is to be remembered from the Ithaca sessions; Coblentz reported on the lunar temperature, which in another abstract in the next year he finds to have been in error.

PHYSICAL REVIEW volumes are getting thicker; even with deficits decreasing, Cornell was still losing money in its venture, this in spite of a growing number of physicists and thus subscribers. Volumes XXIV and XXV (1907) each run to 500 pages, and it is not all Physical Society abstracts either. Contributed paper titles continue generally to be of less interest than those of papers presented at meetings, but their significance seems to be increasing. Ives has another color photography system, mixing the Lippman with the Joly scheme (standing waves with screen of tiny filters). He for long would work prolifically in optics; later, at Bell Laboratories, would become a pioneer in long distance television (scanning disc) wire transmission, and the first (1938, with Stillwell) to demonstrate the transverse Doppler effect of Einstein, that piece to be reported in the *Journal of the Optical Society of America*. There was more in our pages from Pierce, including a paper on the use of carborundum in the rectification of wireless waves. More indication of things to come is a paper by Frances G. Wick, another Cornellian, remembered best at Vassar: "Some Electrical Properties of Silicon; I. Thermoelectric Behavior." Her cast rods were all of an impressive 95% purity. Thirty and forty years later, at considerably higher purity, the material would come to make quite a better rectifier than Pierce's carborundum. Her paper would be followed in the next year by three others, on the resistance, on the Hall effect, and on the EMF of cells in which one electrode is silicon. She opens paper I with the prophetic sentence: "The physical properties of metallic silicon, insofar as they have been investigated, show this substance to be of particular interest." She notes the similarity to carbon and suggests that its position between metal and non-metal may explain some of its characteristics. Abstracts of Physical Society meetings include one by W. G. Cady, later importantly into piezo-electricity, on "The Hissing Point of the Metallic Arc," to be followed a few years later by "Rotations in the Iron Arc"; amusing as to what interests people; but we should smile—some of us undoubtedly provide as much amusement to others in what interests us.

A death of the year (still 1907), and frontispiece portrait, was that of D. B. Brace of Nebraska, seemingly to accompany that of Fitzgerald a year earlier. Brace had in 1904 tried another approach to the ether drift experiment. Nichols, in his REVIEW memoir, refers to Brace's attempt to use the half shade polarizer in his effort to detect change in the ellipticity of light traversing a medium moving through space, to check on the Lorentz-Fitzgerald suggestion that a mechanical change in the apparatus caused the negative Michelson-Morley result; if that was it, Brace's refracting medium might show a change. With his sensitive half shade device and the long path (in water), Brace could have detected a difference in velocity (or index) in the two directions of a part in 10^{12}. If Lorentz and Fitzgerald were correct, an effect 10^3 times as great would have been seen. The ether was a puzzle but Einstein was still a year off when Brace did his experiment, published in the *Philosophical Magazine* and duly summarized for APS REVIEW subscribers in *Science Abstracts* for 1905, Abstract Number 2170. In one of the REVIEWs for the year, Cornell's J. S. Shearer reviews three famous books in optics: Drude in a posthumous revision, Schuster, and R. W. Wood. The first two are rather theoretical and Shearer contrasts Wood's book with them, its emphasis being definitely experimental, as all should know. "The book is an inspiration to students and to teachers and will be a great aid in rescuing physical optics from the absurd mathematical symbolism which sometimes seems to throttle progress in this fruitful field of investigation." He should have heard from Webster on that comment, and would certainly have to modify his opinion today. It is indeed a fruitful field, but the mathematical aspects are in large measure responsible for the resurgence in optics. Theory and experiment go hand in hand together, of course, and optics has come a long way since Shearer's day, theory probably leading the way.

Support for the REVIEW by Cornell physicists continued unflagging. Kevles describes the editors as conducting the journal "with a seeming Baconian eagerness to publish the report of any experiment (and then some!) with a disproportionately large number of articles in the REVIEW continuing to be published by Cornell Ph.D.s and faculty." He indicates that about 35% of the articles in this early period were by Cornell Ph.D.s and Cornell faculty who accounted for 21% of the authors. This presumably covers Series I of the journal. Following its assumption by the Physical Society, participation by "Cornellians" gradually took on more reasonable proportions. A cursory look (in contrast, Kevles did real research on the matter) at the Cumulative Index (1920) for the first Series shows the founder, Nichols, to have had by far the most entries, 143, followed by Merritt with 71, and Bedell with 42. Child, Coblentz, Franklin, and Rosa, all with Cornell roots, came in with 26, 33, 33, and 26, respectively. Not every index entry was for scientific work; book reviews, obituaries, and notes were included. Still, Cornellians were surely in there helping make it go. As we have seen, much of the good American work, little enough of that, went abroad. Nor did the Astrophysical Journal help matters for the REVIEW. But things were improving, work gradually becoming more substantial, as it is hoped the foregoing record indicates. Kevles also quotes Carl Barus (who had 26 Index entries) as pointing out in 1903 that "most of our physicists are young men still under the influence of their thesis work or of ideas derived from somebody else." But those going to Europe were younger men and it was their exposure to new ideas which would eventually raise American physics from its pedestrian level.

In 1908 with Volumes XXVI and XXVII, we find some of more "modern"feel, still more in the APS Abstracts than in Contributed Papers. Millikan reports on the charge carried by "negative ions" of an ionized gas in the cloud chamber method of H. A. (not C.T.R.) Wilson; 4.03×10^{-10} esu is reported. Lyman has an abstract on short wavelength radiations; "his" series in hydrogen is not yet manifest. And Daniel F. Comstock (MIT) has one on "The Electromagnetic Momentum of a General Electric System," neither G.E. Co. nor Einstein. O. W. Richardson has come to America for his period (1907–1913) at Princeton and reports on the emission of negative ions from hot metals, work he will continue on to his Nobel Prize. Substantial contributed articles come from S. J. Allen, "On the Range and Total Ionization of the Alpha Particle," from G. B. Pegram and H. W. Webb on the "Heat Developed in a Mass of Thorium Oxide, due to its Radioactivity," from Frances Wick on her silicon measurements, and again from Richardson on more of his work. Barnett, from Tulane, has a long experimental paper on the electric displacement produced in insulators moving in magnetic fields, something of a corollary to a coming paper by Goddard, in 1914, well ahead of us. At the APS winter meeting in Chicago, the REVIEW founder was elected president of the Society. At this meeting also the Proceedings note in the business session Minutes that the Society voted not to accept the invitation to cooperate with the College Entrance Examination Board (news to the writer that it was by then in existence) in establishing a commission to consider revising entrance requirements in physics; "... declines on the ground that the teaching of physics and all pedagogical matters lie outside the province of the

American Physical Society." This was at the time clearly not the opinion of the REVIEW. Book reviews, lecture demonstration papers, papers on units, teaching approaches all, would attest to this fact. Indeed, numerous papers are devoted to instruments not necessarily even given to physics (i.e., "A New Type of Sextant," Volume XXVI). Furthermore, its editors were in the business of teaching. The position of the Society in this regard was of considerable disappointment to Webster; but he apparently kept pressing his views and eventually the Society got interested in teaching (not to mention other political and societal (small s) matters).

With Volume XXVIII, in 1909, "modern" theory receives some note and a first recognition of sorts. In a paper by Jacob Kunz at Illinois, "On the Electron Theory of Thermal Radiation for Small Values of (λT)," he notes Planck's success but considers alternatives in the Lorentz-Thompson picture of "Roentgen pulses." He says of the electron theory: "... h does not appear to have such a simple physical significance as in the theory of Planck. It is evidence, however, that we do not understand this quantity nor the nature of the unit of electromagnetic energy from a pure physical point of view." Amen. Even today is this not still the situation at rock bottom? If REVIEW readers had not heard much from or about Planck in the pages of their journal, some recognized the importance of what they had read in the European journals or had seen abstracted in their *Science Abstracts*. For at the Washington meeting, Planck was voted in as an honorary member of the Physical Society, although he never did report on any of his work on this side of the Atlantic, either at APS meetings or in REVIEW pages. It is strange, in view of the recognition, that so little mention of his contribution comes through in American science reporting of the period. He would in 1927 receive that medal from the Franklin Institute we have mentioned, along with Nichols' own award. Relativity, however, is at last beginning to appear, but so far only in an APS meeting abstract, an important one by R. C. Tolman and G. N. Lewis, then both at MIT, in "Non-Newtonian Mechanics and the Principle of Relativity," given at the 1908 Christmas meetings, held this time in Baltimore. Relativity, albeit a single entry, appears in a volume index for the first time. Clinton Davisson, at Princeton, working under Richardson (and later marrying the latter's sister) submits at the Washington meeting a "Note on Radiation Due to Impact of β-Particles upon Solid Matter," and Percy Bridgman is into his high pressure research at Harvard with three abstracts from the Baltimore sessions. Future Nobelists are publishing in the journal.

In Volume XXIX, also 1909, there is more from Cornell by another woman on another interesting material—selenium. Louise McDowell writes the first of her three papers "On Some Electrical Properties of Selenium: I." II and III appear in successive REVIEW volumes. She would go on to a long-time professorship at Mt. Holyoke, as Miss Wick would likewise do at Vassar. Pierce is still rectifying oscillations at Harvard, one way and another. Orin Tugman, also at Cornell, mentions a very significant development. He is writing on "The Effect of Electrical Oscillations on the Conductivity Imparted to Gases by an Incandescent Cathode." He refers to the Fleming valve (rectifier) and to Lee DeForest's "Audion" and the latter's patent on a "wireless telephone receiver in which there are two metal plates, one on each side of the filament." The day was not far off when control would be obtained by making one plate an open, wire grid, interposed between the filament and the other plate. High technology was on the way. At the Fall APS meeting in Princeton, Richardson reports on "The Reflection of Low Speed Electrons," and G. N. Lewis on "The Lorentz Shortening—An Apparent Paradox," and W. S. Franklin, in a "Title Only" abstract, discusses (presumably) "The Relationship between Entropy and Time," sophisticated enough. Millikan is observing single fog drops under the influence of gravity and an electric field, giving a value for the elementary charge of 4.65×10^{-10} esu; he is closing in. Franklin follows up his "abstract" with a contributed paper in the next year (Volume XXX), "On Entropy." With that and one more paper two years later (on "gyrostatic" action) he evaporates from REVIEW pages, to which, up to this time, he was a very frequent contributor. He was at Iowa for five years before going to Lehigh (1903–1915), and at MIT from 1917 until retirement in 1929; a very prolific writer of texts. He must have been a student of Nichols at Kansas; he finished his B.S. degree there the year Nichols left, and for his advanced degree (D.Sc.) he came to Cornell a year after Nichols did. (A second Kansan who Nichols lured to the East?)

In Volume XXX (1910)—really fat at 798 pages—one finds more relativity with Tolman's abstract on the Second Postulate (c independent of the motion of source and observer) and with G. N. Lewis again on "The Clock as a Measure of Time and Space." Earlier, at the previous APS meeting, MIT's D. F. Comstock postulates in his abstract, "A Neglected Type of Relativity," that if one assumes a "c" dependent on source velocity, then we have a simple type of relativity presenting "striking contrast to the principle now in vogue." Proctor, at Dartmouth, was repeating the experiment of Kaufmann, by now that of Bucherer and of Allen, on the variation of e/m cathode rays with velocity. There was consensus that it varied but disagreement on the functional relationship, but Proctor indicates "... the weight of evidence since the recent publication of Bucherer's results is certainly on the side of Lorentz-Einstein rather than the Abraham theory," in contradiction to Kaufmann's conclusion. G. A. Cline, studying "The Penetrating Radiation at the Surface of the Earth" finds the intensity somehow connected to the atmospheric barometric pressure. Compton (K.T.), at Ohio's Wooster College, makes his appearance with a paper on the Wehnelt electric interrupter, a device to replace the mechanical circuit breaker used to make sparks in early wireless telegraphy. Millikan is studying the photoelectric effect in the UV and Richtmyer is studying the effect in the alkali metals, and applying the effect to laboratory photometry, one of, if not the first to do so. He would become very much involved with X-ray research as Millikan would not. But here in Volume XXX, W. R. Ham has "Polarization of Roentgen Rays" under Millikan's direction, surveying the intensity distribution around an X-ray tube and finding evidence of polarization expected from the sudden deceleration of the incident electrons. A Will Baker of the School of Mining (!) at Queen's University in Ontario has a paper in Volume XXX on "Bound Mass and the Fitzgerald-Lorentz Contraction," with reference to Bucherer but not to Einstein. The great man's ideas, however, are by now very much on other's minds. In Volume XXXI, Tolman (still at MIT with Lewis—in 1909, following their 1908 abstract in the REVIEW, they had seminal papers in the *Proceedings of the American Academy* and in the *Philosophical Magazine*) expands the APS abstract into a 15-page landmark article describing experiment to show that c is independent of source velocity, and deriving the Second Postulate from the variation of mass with velocity, supported by Bucherer. Further Einstein influence: Jacob Kunz goes haywire in plots of the maximum kinetic energy of electrons in the photoelectric effect, and "shows" that it depends on the square of the exciting frequency; he had earlier in Volume XXX correctly decided on a linear dependence, even obtaining a fair value for Planck's constant. A number of investigators were trying to settle the exponent, crucial to the Einstein interpretation of the phenomenon. To do it right was not a trivial business. Millikan would get into the act but at the moment he is into "*e*" and has switched from water to oil drops and balances them for hours at a time. Nichols and Merritt are well into their long series of papers on luminescence with their fourteenth "Study in Luminescence," and Herbert Ives (again) has "Further Studies of the Firefly," an earlier report on the fascinating insect appearing with Coblentz in the *Transactions of the Illuminating Society* (!). Barnett supports relativity by finding no motion of the ether in a steady electromagnetic field; he seemingly liked looking for small effects such as was here involved in his experiment, not as noted as the one six years ahead. And another Ives, James E., at Cincinnati, uses the interferometer to get wavelength and logarithmic decrement of a "linear," electric, spark gap oscillator. His oscillator we recognize as a half wave dipole, a gap at the middle across which a spark jumps; his receiver is similar with a sensitive thermo-junction across the gap in combination with a Coblentz galvanometer. Sender and receiver are located at the focii of parabolic cylinder reflectors of sheet zinc. Microwaves—and quite remarkable: wavelengths down to 10.7 cm.

Volume XXXI brings a notable change in REVIEW history: Macmillan is no longer publishing the journal. While no correspondence concerning the termination has been run into, the Volume flyleaf reflects the change in replacing the name of the publishing house at page bottom simply with

THE PHYSICAL REVIEW
Lancaster, Pa. and Ithaca, N.Y.
Berlin: Mayer and Mueller

The change would seem to have made more work for the editors but there is no sign that it did. At least now, manuscripts could go direct between Ithaca and the New Era Press in Lancaster. The financial savings were undoubtedly considerable, as Bedell indicated in that letter to Professor Newcombe, and which were in part responsible for the REVIEW's turning a small profit from here on out, which wouldn't hurt if the Physical Society were someday to consider taking total responsibility, a circumstance that could hardly have failed entering the minds of the three conductors.

In a Proceedings of the APS (53rd meeting) Merritt, as Secretary, notes that *Science Abstracts* has been in the kind of financial bind under which the REVIEW has labored, and it is now raising the price to APS members from $1.50 to $2.50 "a copy" presumably a year of

Abstracts). The annual dues of the Society would have to increase; we had already gone from $5 to $6 annually, a move affected at the 41st meeting of the Society in 1908. The Society Council has authorized the increased payment to *Science Abstracts*, so members would still get copies along with their PHYSICAL REVIEWs. Papers at the 53rd meeting included more from Percy Bridgman, Child, McDowell (her selenium), Ives (Herbert) and Richardson, who had one rather out of his line: "Gravitation and the Electron Theory." The paper occasioned the first Erratum to appear in a REVIEW volume index; two terms are missing in the printing of the prime equation. Physicists of the day must have been very careful; today's REVIEWs and REVIEW *Letters* are replete with corrections.

Chapter 7
A New Sponsor

•◆•

In 1911 the *Philosophical Magazine* (*Science Abstract* Number 1347 for the year) has the great paper by Ernest Rutherford on the nuclear atom. No such tremendous breakthrough yet for the PHYSICAL REVIEW but we do have in Volume XXXII, of that year, the 50-page landmark paper of Millikan on his and Harvey Fletcher's determination of the electron charge, and C. D. Child arrives first at the Child-Langmuir, space charge limited, current equation. Langmuir would get there independently three years later in a REVIEW paper; not all were reading the journal cover to cover. C. C. Trowbridge (Columbia) has a nice paper on the decay of the afterglow in a nitrogen ring discharge, and O. M. Stewart discusses the Second Einstein Postulate as it relates to the electromagnetic emission of light. There is C. V. Raman's "Motion at Nodes of a Vibrating String," followed by a note on photometry by Ives again, and by A. H. Pfund's "New Method of Producing Ripples." The Minutes in the published Proceedings of the Physical Society begin to be of more interest to this story, particularly those of the 1911 Spring Washington meeting (57th), as compiled by Professor Merritt, Secretary.

A portion of the Friday session was devoted to a discussion of relations of the PHYSICAL REVIEW with the Physical Society. Be it remembered that the journal came at a bargain rate to Society members. The Minutes tell of what transpired. To start things off, APS President Magie presented a long communication from the REVIEW editors, who were taking that means because a planned conference on the subject had only drawn four members. Apparently a mail ballot had been taken in consequence, of which the letter told, on the question of the relations between the Society, the REVIEW, and *Science Abstracts*. Many suggestions had come in along with the ballots. These fell into three categories:

1. The Society establish one or more journals (this one from Ames of Johns Hopkins).
2. The Physical Society assume control of the PHYSICAL REVIEW.
3. A representative board of editors be selected to operate with the present editors by giving assistance and advice.

The letter detailed at considerable length some of the pros and cons. The REVIEW editors did not much like Ames' suggestion and clearly favored Option (2), preferably preceded in time by (3). Following the presentation, there was discussion carried on in committee of the whole. Two resolutions were adopted:

1. Resolved, that this meeting request the Council to present to the Society a plan by which the Society shall publish a journal.
2. Resolved, further, that the Council be requested to enter into negotiations with the editors of the PHYSICAL REVIEW to determine under what conditions the control of the REVIEW may be transferred to the Society.

All this business was followed by the usual meeting presentation of a long list of papers: included were Richardson, Davisson, Coblentz, Kunz, D. C. Miller, Merritt (on the silicon detector for short electric waves), Rosa, Dorsey, and Charles F. Brush in, appropriately the last paper of the session, on a "Kinetic Theory of Gravitation," concluding that for a falling body, "the gathered energy comes from the ether through which the body falls," and that "gravitation is a push toward the attracting body and not a pull." Brush was clearly involved with gravity, in more ways than one; there would be others of his papers of similar flavor coming in due course, all presented orally at APS meetings. One imagines that an editor or referee would have disallowed publication of his work as contributed papers to the REVIEW. We have the same situation today.

In the Proceedings of the 58th APS meeting (Volume XXIII, 1911, Fayerweather Hall) there was another brief discussion of journal-Society relations, again in Committee of the whole, and the same two resolutions were adopted by the conferees (why this was done twice in successive meetings is not clear). Bridgman had an abstract on the "Effect of Pressure on the Liquid and Five Solid Forms of Water" and Compton (K. T. again—we haven't seen A. H. yet) discusses the contact potential influence of measured electron velocities in the photoelectric effect, and Richardson continues his way with positive and negative ion emission from hot substances. More important in the year's papers is that of Harvey Fletcher on the "Verification of the Theory of Brownian Motion and a Direct Determination of the Value of Ne for Gaseous Ionization." The juxtaposition of this paper and that of Millikan in the preceding volume is of chief interest for Fletcher's posthumous revelations (*Physics Today*, June, 1982) concerning the collaboration. The photoelectric effect was also beginning to occupy Millikan as it was others. The dependence of maximum initial electron velocity on wavelength was of obvious interest in the light (no pun intended) of Planck's quanta and Einstein's running with the concept. Ladenburg had found a linear dependence between maximum initial velocity squared and the frequency. Richardson and Compton, working in the ultra-violet, will come to agree. The above contact potential paper pertains to their work. J. R. Wright, under Millikan at Chicago, and Kunz at Illinois (Volume XXXIII) did not agree, each finding something quite different, and strange behavior. Poor Wright had his photocell on the pump for six months before sealing it off for the experiment. At that, the vacuum was but 10^{-5} mm Hg; pretty clearly some of the weird results came from surface contamination. Millikan would eventually get to the linear dependence and a recognized value of Planck's constant. But it would be another three or four years.

At the 60th APS meeting (Volume XXXIV, 1912), held at the year's (1911) end in Washington, President Magie reported progress in the "consideration of a plan for the publication and control of the PHYSICAL REVIEW by the Society," but no details on the progress. Two unrelated amendments to the constitution were proposed: past presidents of the Society would go on the Council, and election procedures were specified. There was a long program of papers and a symposium on "Ether Theories" led by Michelson. On the panel were Morley, Webster, Franklin, Comstock, and G. N. Lewis (where was Tolman, who seemed to know the score?). The Treasurer reported receipt from 554 members of dues in the amount of $3,323.71, a footnote pointing out that "one member is in arrears $0.29."

The 61st meeting Minutes in Volume XXXIV tell of the adoption of the above proposed constitutional amendments; and a By-Laws amendment was adopted specifying that "at least four regular meetings each year" need not necessarily be held in New York. But no mention of progress in the matter of the REVIEW; presumably it was being worked on. There were papers, however; Richardson is looking at "Ions from Heated Salts," A. W. Hull, at Worcester Polytechnical Institute, in his first of many appearances yet to come is into "Photoemission in Magnetic Fields"; others also.

The 62nd meeting of the Society was held in Cambridge at the end of April, 1912. The Proceedings (Volume XXXIV still) make no note of any business meeting. But the Minutes of the 63rd meeting (Volume XXXV) indicate that the Council did meet after the formal Society session and adopted a suggestion from the REVIEW editors that a mail ballot be sent to the membership concerning transfer of the journal to the Society. Papers given in the formal session included a significant one by Richardson and Compton on their study of the photoelectric effect making use of ultra-violet light and generally confirming Einstein's equation, an important result, notwithstanding Landenburg's previous work in the visible range. Their full report went to Britain and the *Philosophical Magazine*. An intriguing paper was given by E. L. Chaffee of Harvard, who has an interesting "Direct Measurement of the Velocity of Kathode Rays and the Variation of Mass with Velocity." The method times the flight of his electrons across a fixed distance, making use of high frequency oscillations in a rather curious way. He seems never to have given any results, although he claims it should be good to 4%. (Years later, in Volume 36, 1930, in a nice modification of his method Perry and he would measure e/m fairly precisely, but not its variation with velocity.) Harvard, meeting host, is in there again with Bridgman collapsing thick cylinders under high pressure. G. W. Stewart from Iowa is studying the acoustic shadow behind a rigid sphere, getting into the subject of his well known book on acoustics. His brother (O. M.) at Missouri, is more concerned with relativity and matters electrical and radioactive. Both earned their degrees at Cornell and published frequently in the REVIEW.

The 63rd meeting of the Society learned of the outcome of the Council meeting in Cambridge. We are once again in Fayerweather Hall and, in the absence of both the President and Vice-President, Merritt is both presiding and keeping Minutes, which appear in Volume XXXV, 1912. He reported to the meeting the result of the mail ballot which had been suggested by the editors to the Council. The results of the balloting were presented. There were five propositions or choices in the vote:

Proposition 1:
No change be made in the APS-REVIEW
relationship   87 votes
Proposition 2:
That the APS take over the REVIEW on
January 1, 1913, under conditions specified
in the letter accompanying the ballot   97 votes
Proposition 3:
That the Society forget the REVIEW and
put out its own journal   16 votes
Proposition 4:
That the Society Council be given full
powers as Society representative to obtain
a certificate of incorporation in a suitable
state   129 votes
Proposition 5:
That if the Society votes to assume REVIEW
control, the present editors and advisory
editors shall continue for one year and
board prepare a plan for the journal's
permanent conduct   130 votes

There was a problem with the results. With 316 regular members
who could vote, if proposition (2) were to carry, 159 votes were
necessary. Instead they got only 97 out of the 201 ballots sent back.
The editors had intended that all members, regular and associate
alike, have a voice in the decision. Things were left in a bit of a
muddle, for 97 is not quite half of the 201 cast, nor is it overwhelm-
ingly greater than the 87 for no change. It is not recorded in the
Minutes what was decided beyond an affirmative vote to a motion
that the Council inquire into the practicality of obtaining a collective
guarantee of a number of colleges to make the financial position of
the Society secure if it were to go ahead with the publishing of the
REVIEW. Following this rather unsatisfactory session, conferees
heard about a dozen papers, among them A. W. Hull, on reflected
electrons, Richardson on secondary X-radiation, and Herbert Ives on
"The Intrinsic Brightness of the Glow Worm." This indefatigable
optical experimenter liked those bugs. This was certainly not among
the most significant of his over 100 papers in optics, for which in
1938 he would receive from the Optical Society the fourth Frederick
Ives (his father) medal, awarded biannually, the previous three going

to the REVIEW's founder, E. L. Nichols, Theodore Lyman, and R. W. Wood. (The elder Ives, who died in 1937, was, in the Anthony period, Cornell's photographic technician. Morris Bishop writes of him: "This self educated countryman refused an instructorship, migrated to the Great City and invented the three color printing process, the modern form of the binocular microscope, and, it is claimed, the half tone screen.")

In January 1913, members of the APS and others received the first copy of Series II of the PHYSICAL REVIEW. Still a Journal of Experimental and Theoretical Physics, but now conducted by the American Physical Society with a Board of Editors under F. Bedell as Managing Editor. Ames, Guthe, Mclennen, Magie, Millikan, Nichols (E. F.), Peirce (B. O.), Skinner, and Zeleny (J.) made up the board. Preceding the Contributed Papers section was a short statement announcing the change, signed by A.G.W. Deep appreciation for the preceding editors for their twenty years of service and to Cornell University for assuming the financial risk was expressed. The statement went on: "During nearly twenty years the original editors have carried on the arduous task of maintaining this journal on a high standard and it is difficult to estimate the value of their efforts in furthering the cause of physics in America." (Is he implying here that the contribution is just too great or that the physics would have come along anyway without the journal? As has been stressed above, they surely came along together.) "In this manner the way for the foundation of the American Physical Society was prepared (seems true enough), and early in its history the Society and the REVIEW entered into relations that have continually become closer."

The Minutes of the 65th APS meeting, New Year's, in Cleveland, tell the story. They appear in this Volume I of the new Series. There was the election of officers (B. O. Peirce, President, Ernest Merritt, Vice-president, A. D. Cole, Secretary, and J. S. Ames, Treasurer. Strangely enough, Peirce was never a contributor to the REVIEW. He was really a mathematician, Professor of Mathematics and Natural Philosophy at Harvard, and is best known to physicists perhaps for his "Table of Integrals," and for a book "Theory of the Newtonian Potential Function." He apparently did some research on the thermal conductivity of stone, but it never appeared in the REVIEW, where it would have been quite at home.) Following the election there was announcement of the continuation of the arrangement with *Science Abstracts* for another year; a note was taken supporting the rule that previously published work would not be accepted for presentation before the Society; and the Treasurer's report given ($3,468.00 received from 578 members—no one in arrears) showing an increase in bank balance over that of the previous year of $60.28 to $1,077.45 (audited by A. H. Pfund). The business at hand then turned to the PHYSICAL REVIEW. The indecisive vote of the 63rd meeting led the journal editors to re-examine the situation and to send, in a long letter, notice of their evaluation to all APS members. The editors had grave concern over the vote. Since less than a third of regular members favored the change in REVIEW status to that of going over to APS control, the vote was technically a declination of the editors' offer. The vote was not convincing one way or another; while it gave considerable sentiment in favor of the change, "it was not to the overly enthusiastic approval hoped for, as essential to the success of the plan."

From later discussion and correspondence on the matter with various individuals, the editors came to believe that the ballot failure for a clear choice was largely that the editors' views were not understood. They hope their position would be made clear in the long letter that was again sent to the membership. In it they discuss the financial situation; over the previous two years during which the proposed transfer has been on the fire, the journal had moved from deficit to increasing profit. But there would be risks if the Society took it over; expenses would be greater in amount of the editor's salary; but even with that, if managed conservatively and if material does not greatly increase, it ought to be able to get along without dropping Science Abstracts or increasing dues. If, however, scientific activity increases and if the transfer results in articles for the REVIEW which "now go elsewhere—and this result is certainly to be desired ... " the expense of publication would increase. "If the transfer is made," they wrote, "the members should feel that they are taking the REVIEW for better, for worse."

The letter that was sent to members points at the long, difficult, uphill effort in bringing out a journal. The conditions for transfer must safeguard journal interests and APS members' interests. The editors cannot risk losing what has been gained: a journal with an established position, a healthy subscription list, and income of over $3,000. Under the APS it could become even more successful but only with "hearty support" of the members. Unless the REVIEW editors are convinced that members are prepared to take a "vital interest" in the REVIEW success, to transfer would be "little short of a breach of trust."

Assuming the enthusiastic support, however, they are "heartily in favor" of the transfer of the REVIEW to the Society and look on it as in the best interests of both parties. They ask prompt return of the enclosed postcard (prepaid? 1¢!) with preference indicated. The letter was dated December 11, 1912, and signed by the three editors.

The response to the letter was of so large a majority in favor of the transfer that the editors felt justified in approving it. They presented to the AFS Council the proposition which included various technical details: the matter of the Editorial Board, its selection, the non-transferral of office hardware, back volumes on hand remain the property of the "present management," etc.

The APS meeting was informed of the Council's approval and the appointment of F. Bedell as Managing Editor, if the Society went along with the Council. It was moved, seconded, discussed, and voted; the Society went along. There were no dissenting votes. Thus the PHYSICAL REVIEW went over to the American Physical Society "for better, for worse."

Bedell made a motion about his financial bonding and annual auditing, which was okayed. And a vote was taken that a statement be inserted in Number 1 of Volume I of the new Series, II, of the journal starting with the 1913 January issue. A. G. Webster was requested to write it, as he did.

The meeting, on motion of Professor Webster, voted to express its appreciation to Professor Merritt as Secretary of the Society for his services over the thirteen years of the Society's existence.

Besides conducting this auspicious piece of business, members had earlier heard papers in two sessions programmed by the A.A.A.S., Sections A and B, with which the Society was in joint meeting, and in the APS session of its own. Webster spoke to both A and B, Millikan to B and to the APS, as did Dayton C. Miller and L. A. Bauer. Busy men. An eventful meeting.

Chapter 8
SERIES II
THE FIRST DECADE

••

On the shelves of Cornell's Clark Physical Sciences Library, the commencement of Series II of the PHYSICAL REVIEW is noticeable only by a change in the manner of binding covers and the disappearance of gilding on the page edges. Inside each volume, the change is noticed principally by the appearance of advertising, some of which is retained in Cornell's binding: Franz Schmidt (projection equipment), E. H. Sargent and Co. (Bates Polariscope), Adam Hilger (a new ultra-violet monochromator—used by Compton and Richardson in their work on the Einstein photoelectric equation), Leeds and Northrup, C. H. Stoelting Company (Chicago—their nifty Number 2683 Receiving Station of No. 2681 Wireless Outfit pictured), the Cornell Co-operative Society (send for Cross Section Sample book), Weston Electric Instruments, Inc. (where went Western of Volume V?), Leybold's Nachfolger, General Electric Company, Central Scientific Company, Gaertner, J. G. Biddle, Societe Genevoise, Gurley Instruments (good optics from Troy, N.Y.), and others; not book publishers. Mostly familiar names still today. The use of advertising to bolster revenue was continued up to the time that the American Institute of Physics took over the publication twenty years ahead, in 1933. After that, the advertising appeared in the *Review of Scientific Instruments*, in 1933 newly started, and in the *Journal of Applied Physics* (from its inception through 1970).

In 1913 the July issue of the *Philosophical Magazine* carried a truly landmark paper, "Constitution of Atoms and Molecules" by N. Bohr, digested for REVIEW subscribers in three pages of *Science Abstracts*, Number 1930 of the year. For the REVIEW, the new Series starts off less significantly. R. D. Carmichael of Indiana has two sound papers on relativity, his feet on the ground: mass, force, and energy in relativity and one on the philosophical aspects of the theory. Verification of the photoelectric equation of Einstein is still open to question by some: Cornelius and Kunz still find the maximum energy proportional to frequency squared, while later in the volume, in another paper, Compton criticizes the work, pointing out that he and Richardson have reported to the APS a linear dependence using ultraviolet radiation, their results already published in full in the *Philosophical Magazine*. An abstract of a paper given by Irving Langmuir at the 66th APS meeting describes a new vacuum gauge—his well known viscosity gauge—which he sees as able to measure vacua down to 10^{-7} mm Hg. Together with his diffusion pump, yet to come, his adsorption measurements, and such, he was making really significant contributions to vacuum technology, essential to the furtherence of atomic and molecular physics. Viscosity of gases was of concern also at Chicago, where Gilchrist under Millikan was making an absolute air viscosity measurement—a worry to the value of "*e*" right up through the thirties and early forties. The straight-forward, "tried and true" method: the twist produced by a rotating cylinder on an inner, closely spaced, fiber suspended, co-axial cylinder.

The next volume, II (1913), in the Series was something of a mile-stone in the journal development, with the publication of three papers, any one of which would mark growing maturity. W. D. Coolidge has a paper on "A Powerful Roentgen Ray Tube with a Pure Electron Discharge," a hard vacuum tube incorporating as electron source a hot tungsten filament rather than the previously utilized gaseous ionization. Attainment of the requisite high vacuum (a few times 10^{-5} mm Hg) was a major hurdle in the development. Langmuir's famous pump was still a couple of years away so that the slow Gaede rotary mercury pump or the Gaede faster (vacuum-wise), high speed (rpm-wise) rotary molecular pump was used in the evacuation, along with lengthy high temperature outgassing proce-dures. It all seems fairly trivial today; seventy-five years ago it was anything but that. Another important paper was that of Langmuir on "The Effect of Space Charge and Residual Gas on High Vacuum Thermionic Current," acknowledging Child's previous derivation of the 3/2 power law but going considerably further with it himself. And there is the famous summary paper of Millikan on the electron charge (and Avogadro's number), arriving at the definitive value of 4.774×10^{-10} esu, which many of us memorized as graduate students and which value held almost up to the 40's when Bearden's notable measurement finally settled the viscosity of air, to bring the oil drop value up a bit to agree with the value obtained from the X-ray method in the precision determination of lattice constants. Students who have read Millikan's book *The Electron* ("Plus and Minus" would be added later), will recognize in the REVIEW article some of the cuts, plots and tables found in his book. In the same volume of the journal, Langmuir had another paper, itself not insignificant, on the vapor pressure of tungsten. His research activities were definitely in the interests of the General Electric Company, which paid him. He was earning his salary. Of interest to older physicists among the profession is the list of APS membership printed in the volume, regulars and associates. Today the regular members would be fellows and the associates would be the regulars. In the list were the names of six honorary members of the Society: Arrhenius, Bjerknes, Bragg, Lorentz, Planck, and Rutherford. A few years later, in Volume V, the whole list is printed all over again, and Weichert joins the honoraries.

So American physics was gaining in significance. In Volume III (1914) we may not have published the Franck–Hertz experiment, which in the year bolsters Bohr's concept, but we did have Percy Bridgman in two long papers in consecutive REVIEW issues on "The Change of Phase under Pressure" (of eleven substances), and Kadesch at the 69th APS meeting (Chicago) makes the first report on Millikan's "machine shop in a vacuum" for work on the Einstein photoelectric equation and the determination of "h", in "The Positive Potential in the Photoelectric Effect." At this meeting there was a symposium on the Quantum Theory with talks given by Mendenhall (—and Radiation), by Millikan (—and the Photoelectric Effect), and Max Mason (—and Statistical Mechanics), by Jacob Kunz (—and Atomic Structure), and by A. C. Lunn (—and Specific Heats), each presentation followed by a twenty-five minute general discussion. A long afternoon. Kadesch has a contributed paper in the volume, "The Energy of Photoelectrons from Sodium and Potassium as a Function of Frequency of the Incident Light," arriving at an "h" value of about 6.1×10^{-27} erg seconds; and Tolman has another, constructing a miniature universe with a "Principle of Similitude," scaling fundamental quantities in a way such that physics could not know the difference.

Volume III carries the Proceedings of the large, 72nd, Spring APS meeting in Washington; it is noted that a committee was set up to express the loss felt by the Society in the death of B. O. Peirce, reported earlier in the volume. More important, however, were the many papers given at the meetings, among them Lyman's finding in the vacuum ultra-violet two members of a "diffuse" series in hydrogen predicted by Ritz, which "Bears a simple relation to Balmer's formula." Millikan tells of his "direct determination" of "h", and A. W. Hull is reflecting and scattering slow moving electrons. The meetings were unusually large; fully 150 members registered. For Sir (at Manchester, now, he has made his mark) Ernest Rutherford's talk on "X-ray and Gamma Ray Spectra," over

300 people attended, APS ranks being augmented by many of the Electrophysics Committee of the A.I.E.E., in joint meeting with the Society.

In Volume IV (1914), C. V. Raman has a paper on mechanical vibrations (working in acoustics, he reports now and then on such matters), somewhat removed from his later, much more widely known, light scattering. Buckingham introduces dimensional analysis in "On Physically Similar Systems: Illustrations of the Use of Dimensional Equations." And as earlier, Harvey Fletcher follows up Millikan's paper on "e" with another of his own, from Brigham Young University, on the Brownian motion of suspended oil drops and the determination of Avogadro's number. At reduced pressure, he or his student, Carl Eyring, could keep a drop under observation for up to six hours with no change in charge. Impressive to anyone who has worked with oil drops. And a surprise: Robert Goddard, of rocket fame, attempts to measure the force on a dielectric produced by a magnetic field acting on a displacement current set up in the medium; because of ill-health, the experiment is not extended but a positive result is indicated. At G.E., Saul Dushman, collaborator of Langmuir's, obtains e/m through use of the Child-Langmuir equation and space charge limited currents. And O. E. Buckley, one day to head Bell Laboratories, investigates at Cornell, under Merritt, the Hall effect in silicon, following up the earlier work of Miss Wick.

An enlarged section now appears regularly in the journal, one which had almost been dispensed with in Series I: *New Books*. Among the several reviewed in the first of their reappearance, Volume IV, were "Vorlesungen Uber die Theorie der Warmstrahlung" by Max Planck, "Rays of Positive Charge and Their Application to Chemical Analysis" by J. J. Thompson, and Soddy's "The Chemistry of the Radio-elements." In a later issue, Jeans' brief "Report on Radiation and the Quantum Theory" is reviewed; in the book Jeans reluctantly concedes the necessity for involving non-Newtonian laws in governing radiation.

In the next volume (V, 1915) Carl Eyring under Fletcher continues the Brownian motion work, now in hydrogen at reduced pressure in the oil drop apparatus. Saul Dushman writes on the theory and use of the molecular vacuum gauge, and another Langmuir collaborator (at another time—in an "octet" atomic theory based on a cubic model), G. N. Lewis again, now at Berkeley, contrasts the Maxwell distribution law in Newtonian and non-Newtonian mechanics. Raman is in there again and both Comptons appear, separately: K. T. with Trousdale at Reed College on an X-ray diffraction study of magnetized *versus* unmagnetized crystals, "The Nature of the Ultimate Magnetic Particle," in no case discovering that "magnetization influenced the diffraction pattern," concluding that the results show "the ultimate magnetic particles must be atoms or something within atoms and are therefore consistent with the electron theory of magnetism." A. H., from Princeton, has a rather curious experimental paper determining latitude and azimuth of his apparatus axes, and earth rotation period, from a little known effect, to wit: if at the North Pole, for example, one has a horizontal, hollow tube, closed in a circle and filled with water (it's a warm day at the Pole) the liquid comes to partake of the same earth rotation as the tube holding it. Flip the tube over 180° and the liquid and tube rotations are initially opposite. Based on microscopic flow observations of such an arrangement at Princeton, Compton determines the astronomical elements of the apparatus to a surprising degree of agreement with the known values, acknowledging the aid of astronomer Henry Norris Russell. Less successful was his theory in an abstract given in the same volume; "agglomeration," a theory to account for the specific heat variation with temperature; he is looking for (and "finds") a satisfactory explanation which, as he elaborates in his full paper in Volume VI, does "not involve the conception of quanta." "If the relative energy between two neighboring atoms in a solid falls below a certain critical value, the two atoms become agglomerated so the degree of freedom between them vanishes—" and reappears if we back up and increase the energy. The theory provided a better fit to the data than either that of Einstein or Debye, he claimed. It is ironic that he would in a few years nail down the concept of quanta, if the photoelectric work of his brother and Richardson did not.

Illustrative of the growing depth and interest of American physics was the symposium on "Spectroscopic Evidence Regarding Atomic Structure" held at the 74th APS Chicago meeting: H. Lemon on the Nicholsen atom, H. Gale on the Ritz theory, G. S. Fulcher on the Stark effect, G. W. Stewart on "Energy Relations in Light Excitation by Impact," and K. K. Darrow (his first showing in Physical Society annals) on "X-ray Spectra," upstaged four hours later by Professor W. H. Bragg, who gave an evening lecture in Kent Theater on "X-ray Spectra and the Nature of X-rays" (with slides and models); a reception followed at the Quadrangle Club. More significant was the work of Barnett that he reported in an abstract on "Magnetization by Rotation," his full paper following later in the same volume (VI) and a last paper on the work in Volume X, where he uses a slightly different detector system. One recalls the effect, first looked for unsuccessfully by Maxwell: Consider a cylinder of iron. "If it is given an angular acceleration about its axis, each individual system, which we may suppose for simplicity to consist of a number of electrons revolving in fixed orbits with constant average velocities about an oppositely charged nucleus, will change its orientation in such a way as to contribute a minute angular momentum and therefore a minute magnetic moment parallel to the axis of the cylinder," quoting from Barnett's paper. There are a number of similar, small electron inertia effects. Tolman and Stewart would go after another one in Volume VIII.

It is of interest that so much work is first reported in the REVIEW as an abstract of an oral paper given at an APS meeting, to be followed later by a full blown paper as a contributed article in the journal (or elsewhere). In his biography of I. I. Rabi, John Rigden tells of Rabi's learning the ropes. His first contribution to the REVIEW was a full paper on some magnetic susceptibility measurements he had made—in a clever departure from the usual technique to be cited later herein as a REVIEW contribution. His paper was almost unnoticed. Thereafter he always reported first in a meeting, the abstract appearing in the journal, catching readers' eyes, to be followed in due course with the details in his contributed article; a sort of "Coming Attractions" ploy.

Immediately preceding Barnett's important abstract above, there was an even more important one. No mere "Coming Attraction" this; it was almost a full paper—5 1/2 pages, tables, curves, the works. From Harvard, this is the Duane and Hunt famous work, "On X-ray Wavelengths," a title which rather hides its significance. In it they establish a definite short wavelength cut-off in the continuous X-ray spectrum (Coolidge's tube) and make a precision determination of "h". It is perhaps odd that no further work on it is reported as a contributed paper anywhere in our journal. Certainly the "abstract" is convincing and complete enough. The two abstracts alone, that of Barnett and that of Duane and Hunt would make Volume VI noteworthy. Contributing further to the stature of the volume were two more, long, consecutive issue papers of Bridgman again on "Change of Phase with Pressure"—melting curves again. Harvard was doing well by physics and the REVIEW, whatever Franklin had felt about the place twenty years before. In the same volume, Tolman, now at Berkeley with Lewis, applies his principles of similitude and dimensional homogeneity to the deduction of some physical laws—the proportionality between mass and energy; the relation between energy density in a hohlraum and temperature, arriving at Stefan's law; and the gas law, $pv = NRT$, given acceptance of Boyle's law and Avogadro's hypothesis that equal volumes of gases at the same p and T have the same number of molecules.

By this time Europe was well in the throes of World War I. It would not be apparent from papers in our journal—the volumes continue to expand, thickness-wise and serious content-wise. The British *Philosophical Magazine* in contrast, diminished considerably both ways in content. While it was not a physicists' war as would be the next one, the ranks of European physicists were nonetheless depleted for service elsewhere; one recalls especially the loss at Gallipoli of the youthful Moseley. Some Minutes of APS meetings reflected the changed situation over there. Those for the 80th meeting (Volume VII, 1916) make note of the REVIEW's Managing Editor asking what policy should be adopted by the Board of Editors in "attempting to provide an adequate channel for publication." The Society voted its approval of the way the journal was being conducted "thus far" and recorded its "feeling that the Board should feel authorized to make such enlargements in the REVIEW as, in the opinion of the Board, should seem desirable and expedient." At the next meeting the Minutes note that amendments to the Constitution were adopted which were designed to incorporate into the Constitution the more important features of the adoption of the PHYSICAL REVIEW as the organ of the Society. On recommendation of the Council that an effort to secure an endowment for the support of the PHYSICAL REVIEW be made, the following motion was adopted: "The American Physical Society declares that the present advance of production in physical research in this country, as well as the actual conditions in Europe, render absolutely necessary the increase of means of publication; that to this end it is essential that an ample endowment be secured and that a committee be appointed by the President to inaugurate an effort to secure such endowment." (So far as the writer can tell, the motion is still on the books and no endowment as such is recorded as having been secured; the financial support given by many industrial firms and laboratories to the American Institute of Physics, which charges the Physical Society for services rendered, can hardly be counted as Society endowment.) At this meeting the Managing Editor presented his report, which was "accepted and placed on file. It showed an excess of receipts over expenditures for the year of $984.28". Quite a change from the red of a decade before. The membership numbered 703.

In the same REVIEW volume (VII) there appears an article by Millikan: his definitive last word on "h" with his "vacuum machine shop," which allowed him periodically (not in the Fourier sense) to scrape a fresh surface on his emitting material. He was surely on the way to his Nobel award with the two important works on "e" and "h", together with his other contributions. A. H. Compton is on *his* way publishing more often in X-ray physics than, say, in geodesy or "agglomeration," his "Recording X-ray Spectrometer and the Spectrum of Tungsten" appearing during the year. And Langmuir at G.E. is on *his* Nobel path; in Volume VII offering three papers on various manifestations of tungsten, but one interest in the many-sided and significant researches on which he frequently reports. A. W. Hull, now also at G.E., is looking at the reflection of slow electrons, from copper; that should be Clinton Davisson and the material nickel, but they will come. Rather, not yet at Western Electric (Bell) Laboratories, Davisson is at Carnegie Tech worrying about the dispersion of hydrogen and helium on the basis of Bohr's theory, and his results are pretty unsatisfactory; but, at least, Bohr's concept is being worked on. Hull is beginning to be seen more and more in the journal. He is an imaginative physicist, suprisingly relatively unheralded, from whom one comes away with great respect in his papers. At the 79th APS meeting he was demonstrating the negative resistance in a vacuum tube occasioned by a secondary emission coefficient greater than unity, and in another paper given at the 80th meeting he confirms the Duane-Hunt limit up to 100 kV, a result which disagreed with that of Rutherford, which showed the frequency limit to increase quite less rapidly than linearly. At that same meeting, in Chicago, Michelson "at home" departs from his usual interest in a paper on "The Laws of Elastic and Viscous Flow." No abstract subsequentially appeared, but one might surmise that he would no doubt work optics into it some way, as he would four years later with Gale in the tidal measurement of the rigidity of the earth. Curiously, that surmise is wrong: he published the work on the laws of flow in the 1917 *National Academy Proceedings*; in it he only mentions light in describing the weight of a small pulley. Another curious paper: the eventual grand old man of the Physical Society, K. K. Darrow, is not generally regarded as an experimental physicist. Yet at the 81st meeting of the Society, he reports his Chicago experiment "On the Velocity of Sound and the Ratio of Specific Heats for Hydrogen," and his REVIEW article would follow in due course (Volume VIII). At the same meeting, A. G. Webster extended an electrical concept with a report on "Mechanical and Acoustical Impedance and the Theory of the Phonograph"; no abstract or subsequent paper ever appeared in the REVIEW.

A future editor of the REVIEW makes his bow in Volume VIII (1916) in work with which he would long be connected. The APS abstract (83rd meeting) is by John Tate on "The Low Potential Discharge of Mercury Vapor in Relation to Ionization Potentials." While he reports from the physics laboratory at Nebraska, named by then for the aforementioned and deceased D. B. Brace, Tate would in the minds of most be remembered as associated with the Physics Department at Minnesota. At this same 83rd meeting of the Society, Langmuir revealed his "High Vacuum Mercury Vapor Pump of Extreme Speed." His full paper on it comes later in the volume.

Well known names (at least to the elder among our ranks) appear more and more frequently in REVIEW pages and the significance of the work grows. We have Millikan, the Comptons, Ives, Bridgeman, Langmuir, Hull, Coblentz, Foote, Forsythe, Kunz, Websters (David and A. G.), many more we know. Not that there are no forgettable papers. One such, rather amusing, in Volume VII, is by one Alfred J. Lotka; if you can't find the lens to your photographic enlarger, he has the answer—forget it. He describes a lenseless photographic enlarging process he has patented. A negative is moved horizontally in the X-direction under an illuminated narrow Y-direction slit. The plate emulsion being exposed is moved beneath the negative in the X-direction but at a faster rate. Thus a horizontally distorted positive transparency results. Use a negative contact print of this then and repeat the process in a vertical Y-direction scan with horizontal X-direction slit. His reproduction of a cow looks quite like a cow. The process is akin to turning up the vertical and horizontal gain controls of an oscilloscope. But we have to be careful about whom we are taking somewhat lightly around here. A recently noticed reference to Alfred Lotka in the *Smithsonian Journal* (May 1988, on a conference about the "Living Earth") months after writing the foregoing, sent the writer to the *World's Who's Who* ("from Antiquity to the Present"). There he is; mathematician, statistician (Metropolitan Life) and a very sizable and respectable entry, including the fact that he earned a Master's degree from the Physics Department at Cornell (1908–1909), serving as assistant the while. Indeed, he appears for that year in the Physics wall chart of the Department characters ("from antiquity to 1955," when it was discontinued). It is clear that Lotka's reputation rests less on his work as photographer than as statistician, in which area he came to formulate Lotka's Law of Productivity, this which states that the number of workers producing just N papers is proportional to $1/N^2$. The Law is apparently well recognized and is followed remarkably well from earliest publications to the present. Price refers to it in his "Science Since Babylon," previously cited.

Nevertheless, the contrast between the journal content of these pre-1920 years and what it had started with is remarkable. And the number of published pages continues to grow; Volume VIII carried 750 pages and includes (besides the previously cited Davisson *versus* Bohr, in dispersion) Langmuir on his great diffusion pump (and before the end of the year, an advertisement at the back of a REVIEW tells that for $90 J. G. Biddle will sell you an all metal version); Tolman and T. D. Stewart on the EMF produced in an accelerated metal because of the electrons' inertia; F. Schwers, from London, knocking down Compton's agglomeration theory, finding "Debije's" theory a much better fit to the data; our ladies, McDowell and Wick, together on the law of response of Merritt's silicon detector for short (1 meter) radio waves; Millikan in his scrap with Ehrenhaft over the latter's sub-electron; Webster (D. L.) on the non-refraction ("no trace" of deviation) of X-rays; and Harrington checking the viscosity of air for Millikan again (there is clearly uneasiness over that quantity) but arriving at the same value that Millikan had used in his determination of *"e."* (Mrs.) Ehrenfest-Afanassjewa from Leiden in a later interesting article, considers Tolman's new principles with skepticism, and Bridgman in a later interesting article, much concerned over the "similitude" implications, considers it also, at some length, and finds in it nothing not already in the theory of dimensions.

In 1917 the United States entered the war to end wars, but there is little evidence of the fact in the REVIEW Volumes IX and X of that year print nearly 600 and 800 pages respectively. Two papers based on Moseley's work appear: Uhler on the "Law of X-ray Spectra" and Fernando Sanford, who still contributes frequently from Stanford, on "The Nuclear Charge of Atoms;" and A. H. Compton is edging further toward his great discovery with a paper on "The Intensity of X-ray Reflection and the Distribution of Electrons in Atoms;" X-rays are going strong. At G.E., with Coolidge's tube, Wheeler Davey gets into biophysics with "The Effect of X-rays on the Length of Life of Tribolum Confuson"—the flour weevil. Later in the year, Barnett has the further long paper on "his" effect, and Davis and Goucher describe their well-known work on the ionization and excitation potentials of hydrogen and mercury; and Hull again demonstrates his versatility with "A New Method of X-ray Quantitative Analysis"—we know it as powder diffraction. He reported it first in Volume IX at an APS meeting in 1916, October at Nela Park. In Volume X, E. C. Wente (AT&T Laboratories) invents radio's condenser microphone ("transmitter") as a means for measuring absolute sound intensity, not for the broadcast of entertainment.

However few, there were nonetheless specific hints and references to the trouble in Europe and our recent involvement. At the 89th APS Spring meeting (Bureau of Standards, Volume X), "President Millikan outlined to the Society the plans of the National Research Council (established a year earlier) for enabling the United States Government to utilize the research ability of this and other scientific organizations for the national defense. Considerable discussion followed." At the 95th, December 1918 (Baltimore) meeting, a month after the Armistice, the Minutes recorded that Vice-president Ames presided, Professor Bumstead having resigned and in London as Military Attache at the American Embassy. "As a matter of record," Secretary Dayton C. Miller notes, "because of conditions growing out of the war, three meetings, which following custom, would have been held in the year 1918, were canceled."

At this 95th meeting, it was by then Colonel Millikan who was explaining the purpose of the Smith-Howard Act, the intent of which was to set Federal co-operation with the states for promoting engineering and industrial research. Two of the 935 total APS membership had died: Ernest Weide, killed in action in France, and C. C. Trowbridge of Princeton of natural cause—his posthumous APS abstract with Mabel Weibel on the "Thermal Expansion of Living Tree Trunks" (!) appeared later in the same volume. His interests more often had been in aurora, meteor trains, and air glow phosphorescence. During the war, other physicists than Millikan took on military officer status: Capt. Lyman and Lt. Col. Agustus Trowbridge (no obvious relation to C. C.), together actively involved at the war front in sound ranging for the locating of enemy artillery; Lt. Col. Shearer, Cornell's radiologist in charge of American Army overseas radiology; Army Major Ives in aerial photography; Lt. Tate in the Signal Corps; and others left their posts for work at defense and war establishments; still others would work at military stations as civilians (Richtmyer as radio engineer with the Signal Corps, to become a major after the war in the Army Reserve; the Comptons also with the Signal Corps; Merritt and Mason at New London, and Langmuir at Nahant, these three in submarine location). Gun ranging-locating and the detection of submarines were the overriding concerns. Col. Millikan was all over, based in New York at what would be Bell Laboratories some day, coordinating work in the submarine business. George Ellery Hale was the big man, protagonist for science and its involvement, arguing for preparedness, vocal, active, and very largely responsible for the establishment of the National Research Council. Michelson was working on range finders, R. W. Wood on a UV signaling system (and considering trained dolphins as submarine locators); others stayed home giving courses related to the war; in addition to his editorial work, Bedell gave a Cornell course on the airplane, a book or two coming out of it.

In any event there was surely little note of war activity reflected in REVIEW pages or APS Minutes therein. That war work was going on was indicated, however, by the APS Proceedings of the 90th meeting, in Rochester (October, 1917, Volume XI), for instance. ("A cordial vote of thanks was extended to the Bausch and Lomb Optical Company, the Taylor Instrument Company, and the Eastman Kodak Company, and the efficient committee—, etc."). Considering the circumstances, national and local, it is not surprising to find papers on "Submarine Periscopes," "Optical Range Finders for Military Purposes," "Apparatus for Testing Searchlight Mirrors," "On Aneroid Barometers," "Heat Treatment of Mercurial Thermometers," "Images on Silver Photo-plate," "Photographic Sensitometry" (L. A. Jones), "Resolving Power" (F. E. Ross and Kenneth Huse—Ross with Calvert later did a beautiful photographic atlas of the Milky Way with a lens of his design), and a talk by C.E.K. Mees on work being done at Eastman Kodak. The names cited are familiar to many older physicists.

And that's how one might infer at the time that there was a war in progress. To learn of the involvement of American physics and physicists in the war, one can hardly do better than to read the relevant sections in Kevles—all very interesting.

After the war, there came an abundant clue in the REVIEW that something had indeed been going on. The Spring 1919 Washington meeting (the 97th) was full of war related work. Bumstead was back and presided with Ames in some of the four sessions. It was an important meeting; 300 participants, 52 papers read, and 188 given by Title Only, most of them secret stuff. "It is doubtful," D. C. Miller's Minutes note, "whether such an amount of important scientific work has ever before been presented at any scientific meeting in this country." We can tell we had been in a war; papers read include such as "Method of Testing Drop Bombs," "Determination of Position of Shell Bursts by Photography at Night with Stars" (?), "Location of Aircraft by Sound" (our friend G. W. Stewart), "Principles of Camouflage," "Submarine Detection Problems" (Mason of Wisconsin), "Pressure Waves due to Discharge of Large Guns," "Infra-red Signaling" (Coblentz), "High Velocity Jets of Gas, Applicable to Rockets" (Goddard, naturally), "Airplane Photography" (Ives), etc. The listing of the 188 papers by Title Only bristles with weaponry and countermeasures. War aplenty.

The important contributed REVIEW paper of the year 1917 (Volume XI) is probably that of A. J. Dempster on "A New Method of Positive Ray Analysis"—his important 180° focusing mass spectrograph. A New Book is reviewed by Editorial Board member E. P. Lewis, commencing with: "Occasionally in science as in other fields of human activity, a classic appears, that is to say a work which is practically finished and which has permanent value." He is referring to Millikan's *The Electron*. (The University of Chicago Press, Pp. xii + 268. Price $1.50 net. Try that today!) An experimental paper comes from Italy by Q. Majorana on the constancy of the velocity of light reflected from a moving mirror (actually, successively from *ten* stationary mirrors and a like number of moving mirrors mounted on the rim of a wheel turning at 80 r.p.s., 0.7 fringe shift in the light analyzed by a Michelson interferometer observed upon reversal of rotation of the wheel). Leigh Page of Yale skirts today's quantum electrodynamics wondering "Is a Moving Mass Retarded by the Reaction to its own Radiation." The Bohr atom is still of interest and the best atomic theory around; Kunz applies it to Bohr's atom and magnetism.

Papers from abroad are becoming more frequent; contributions by physicists from India seem particularly to stand out. One recognizes Megh N. Saha; his "On the Limit of Interference in the Fabry-Perot Interferometer" (Volume X) and "Properties of the Electron"

(Volume XIII) give no hint of his later work on stellar atmospheres and high temperature ionization. We have already spotted C. V. Raman. There are others less well known. Did they find the REVIEW an easier outlet for their work than the British journals? Saha would later (1920–1921) publish his best known work in Britain, and Raman would publish the account of his light scattering in "his" *Indian Journal of Physics*, earning him his Nobel award.

The vacuum tube amplifier had arrived in time for war applications; it would be the subject of numerous papers to come in the REVIEW. Already, in Volume XII (1918) van der Bijl has "The Theory of the Thermionic Amplifier," a precursor to his well-known book— one of the first but now very much outdated and relegated to library back rooms; the vacuum tube itself has today almost become an antique. In Volume XIII, H. W. Nichols, (another one! There are still others down the road to come but we'll steer clear of any more) also at Western Electric, has another vacuum tube paper on "The Audion as a Circuit Element." In spite of the imminent application of vacuum tubes to measurement in physics, an important paper by the two Comptons appeared after the war, in Volume XIV: "A Sensitive Modification of the Quadrant Electrometer: Its Theory and Use." The usual form of the instrument is modified by tilting the needle and incorporating a displaceable quadrant, edging the arrangement close to instability. The modified form would reign supreme in the domain of low current measurement until the late twenties or early thirties, when it was learned how to increase the input resistance of the vacuum tube, which in its turn would give way to vibrating reed electrometers.

The measurement of resonance potentials of atoms and molecules was popular and coming to the fore: Davis and Goucher; Foote, Rongley, and Mohler; and H. D. Smyth (Volume XIII—nitrogen; XIII—arsenic, rubidium and cesium; XIV—nitrogen; respectively). The beautiful thing about Smyth's paper in contrast to the other two is that an abstract precedes the main article. Indeed, most papers of the volume are preceded by a "synopsis" and the practice would be almost without exception by Volume XVI. There was "Information for Contributors" printed in the back of issues in Volume XII mandating a synopsis to accompany each article—an enormous help in perusal of articles' content. In Volume XV, the Minutes of Proceedings of the 103rd APS meeting record that the Council voted the desirability for meetings of sending in paper abstracts along with the titles in advance of the meeting where the paper is to be presented. A committee was appointed to formulate a practical plan for doing so; it wouldn't appear to be all that much of a problem. It was another step toward making it easier going for the reader; why it took so long to implement these obvious and simple changes is not clear. Up to this point, the reader of the present chronicle has not been made aware of the confusion that attends the practice prior to this. Previous to this action, the Proceedings of an APS meeting published in the Review carried the titles of the papers presented at the meetings and the name of the author. There then follow a series of paper abstracts, but few of which pertain to that particular meeting; mostly they are abstracts of papers presented at previous meetings, identified in a footnote. In some cases no abstract ever follows, and some may appear in a Proceedings two meetings beyond that in which they were actually presented. And then the fully detailed article may appear some time after that. At least, for a long time before this, the home base of the author of each article and abstract has been identified. No more lost A. B. Porters. At the 105th meeting "The proposal of the Board of Editors to issue a general index to the Physical Review covering issues from 1893–1920 inclusive was approved." The Index came in due course; another invaluable aid, and some contrast between it and the next Cumulative Index—1950!

In Volume XV, W. D. Harkins has "The Nucleii of Atoms and the New Periodic System," and with us again is Fernando Sanford in a weird sort of paper, "Is the Einstein Radiation Factor "h" a Constant?" Even in Einstein's photoelectric equation, we commonly think of "h" as Planck's constant. On the basis of relating electron orbital velocity to emission velocity, he calculates the Moseley wavelengths (up to cerium) assuming h^3/m a constant. But we know "m" varies with velocity; "deduction: the quantum constant "h" varies with the frequency." More important is a paper by Kemble on "The Bohr Theory and the Approximate Harmonics in the Infra-red Spectra of Diatomic Gases" and its note added in proof: "—long delayed copies of *Annalen der Physik* for 1916 (understandable the delay) has put into my hands the independent formulation of the same hypothesis by Sommerfeld together with his brilliant explanation of the complex structure of the lines in the spectra of hydrogen and ionized helium by its aid."

The decade of the twenties was a marvelous period in the development of physics. While most of the significant theory was still coming from Europe and published abroad, American physics was making its mark, perhaps more in experimental work than in theory. But the younger physicists from here were going over to Europe in greater numbers and being exposed to the new ideas, then in vigorous ferment. By the beginning of the decade, Einstein's predicted deflection of starlight by the sun had been confirmed in the eclipse expedition of Eddington's. In 1921 both Einstein and Mme. Curie visited the United States, the latter going home with $100K of radium donated by American women, both trips widely publicized.

For the REVIEW of 1921 (Volume XVII) the eclipse brought forward "An Electromagnetic Theory of Gravitation" by H. A. Wilson of Rice Institute. It is fair to say, however, that to this day gravity has not yet been unified with electromagnetism, or anything else. In the same volume, one finds G.E.'s Dushman and Found investigating Buckley's recently devised ionization vacuum gauge (also independently conceived by G.E.'s man Hull) and, leading that paper, a contribution by A. Marcus from a new laboratory which would come to have a very great impact on society, and which was very largely built on the success of the thermionic vacuum tube. From the Radio Corporation of America's laboratory, Marcus calculates the amplification constant of "Weagant's Thermionic Tube Having the Control on the Outside". Not a very important paper; curious enough, from what was to be an important organization, one to become essentially the fountainhead in the development of radio and particularly television, two and three decades down the road. More significant than Weagant's device was one from the General Electric Laboratory, by Hull again. This was his magnetron, another tube with the control on the outside, considerably less subject to erratic charge build-up on glass walls than must have been Weagant's tube. Besides being a neat way for the determination of e/m for the electron, it would play a vital role in radar development of the next great war, when the circular anode surrounding the axial filament was replaced by a circle of resonant cavities coupled to each other and to an antenna. In the same issue of the REVIEW (Volume XVII) Bridgeman, in a long important article, continues work he reported on back in 1917 (Volume IX) on the effect of high pressure on the resistance of 22 metals; and Richtmyer disposes of Barkla's J-radiation in the sequence of K, L, M, etc. series of X-ray spectra. Duane and others also worked on this. Also disposed of by a British contributor, H. H. Potter, was the earlier disturbing report given at an APS meeting by Mr. Brush, earlier seen in his view of ether gravitation; he reported that the gravitational acceleration for bismuth was only 70% of that for zinc. The comfortable, disposing result of Potter in Volume XIX was that bismuth and brass (and we know what brass is made of) are the same to a part in 50,000, checked with a pendulum. Maybe the copper in the brass just compensated for the effect of the zinc. But then, in Volume XX, H. A. Wilson checked the acceleration for aluminum and bismuth, the latter of which Brush indicated showed the greatest discrepancy. Wilson finds they are equivalent in this regard to a part in a million, in the method of Eotvos. (Fifth force proponents might find the result of interest). Brush had another weird APS meeting paper purporting to show silicates fell slower than other materials; in some way the ether kept them warmer than their environment and "so" they fell differently. Another pendulum experiment (Volume XVII) reported satisfactory behavior, with a unique suspension of a Foucault pendulum only $2\frac{1}{2}$ meters in length, not quite down to that of R. H. Crane's contemporary pendulum of less than $\frac{1}{2}$ meter in length, but not bad. Related in physics was the also earlier paper (Volume XIV) by W. Baker of Ontario, on the "Displacement from Apparent Vertical in Free Fall." It would not be necessary to do the experiment which Hall and Cajori had suggested be carried out in the "great pile" that is the Washington monument. The theory predicts a small southerly displacement along with that to the east. W. G. Cady reports in Volume XVII on "A Piezoelectric Method for Generating Oscillations of Constant Frequency," which with miniaturization allowed by solid state developments in the late part of the century would almost eliminate mechanical clocks and watches. Davisson and Germer are together in Volume XX "On the Work Function of Tungsten" at the Western Electric Laboratories, soon to be Bell. And at the 111th APS meeting (Volume XIX) A. H. Compton teases with his "Secondary Spectrum of X-rays;" the softer scattered component is considered to be fluorescent radiation. In Volume XX (1922), Kennedy proposes "Another Ether Drift Experiment" to "detect an ether drift or to confirm the time transformation between relatively moving systems—," results reported on in detail ten years later with Thorndyke in Volume 42. With Volume XX, the title of papers given at APS meetings disappear from REVIEW issue Indices and appear rather in the published Proceedings for the meeting, along with the author's name and a number. The inclusive abstracts which follow immediately are given the corresponding number and ordered serially, and the author's institution noted. How much more satisfactory all the way around for everyone concerned, most of all the reader. It took so long to get there. Nichols was back with the journal, listed on the Board of Editors and winding up a three-year term. Included also on the Board, listed on the REVIEW cover sheet, was the name of G. S. Fulcher; he was elected for a three-year term at the 1920, Christmas (106th) meeting in Chicago.

Chapter 9
A New Editor
(1923–1926)

❖❖

In 1923, with Volume 21 (we get modern; numbering of volumes in the Roman manner becomes obsolete) a new editor took over the management of the PHYSICAL REVIEW; Bedell (or someone else) decided that he'd had it long enough. During the year, he and his wife took that rugged previously mentioned automobile "highway" exploration around the United States. It is not clear whether it was intended that the replacement editor, G. S. Fulcher, based at the glass works in Corning, New York, by automobile an hour's drive from Ithaca (these days), would serve only an interim term, or not, but the fact is that he managed the journal for but three years, the period for which he was elected Managing Editor at the 118th (1922) regular APS Christmas meeting in Cambridge, the Proceedings published in Volume 21 (which noted also that I. I. Rabi was among those elected to membership). *Who Was Who in America,* Volume 77 (1977–1981) has this to say under "Fulcher, Gordon Scott (Fool'sher), 1884–." He earned his B.S. degree from Northwestern University and took an M.A. and Ph.D. from Clark University in 1906 and 1910 respectively, probably under Webster. Following a year, 1908–1909, as Instructor at Amherst, he was at the University of Wisconsin until 1918, serving during the war in the Bureau of Aircraft Production. The short biography notes that he helped develop a leakproof gasoline tank for airplanes. He was next at the Corning Glass Works as Research Physicist until 1931, winding up his scientific career as Director of Carhart Refractories in Louisville, Kentucky, in 1938. While he is cited as author of "Better Thinking for Better Living," one suspects his Wisconsin research papers are more enduring. Why he was selected as Editor, and after only a two-year stint on the Editorial Board, can only be surmised; he was close to Ithaca, which had been the management base from the time the journal started, and he was known to Webster. The record seems to be missing a few points here. His contributions to the REVIEW appeared only in APS meeting abstracts, work related to "canal rays." He published in the *Astrophysical Journal* and *Zeitschrift für Physik* on the production of light by "canal rays," papers which appear quite respectable and significant. He is listed in APS Proceedings (Volume V, Series II) as on the program of the 74th meeting with a paper on "A Method for Determining the Variation of Frequency of the Light Emitted with the Speed of Source." As we have complained, it is typical of the REVIEW of the time that the abstracts appear most often in a subsequent issue with the footnote that it was presented at such and such earlier meeting. Fulcher's abstract showed up in the next issue but titled considerably differently as "A Method of Determining Whether or not the Velocity of Light Depends Upon the Velocity of the Source, by the Use of Canal Rays," reminding one of the Ives–Stilwell, 1938, landmark paper demonstrating the transverse Doppler effect, also using "canal ray" light.

Fulcher's first REVIEW volume was notable for at least three items, beyond the introduction of the Arabic numbering system: there was the obituary for A.G. Webster, written by G.S.F., A.H. Compton's epochal APS abstract and two papers on his Nobel prize effect, and the paper of Nichols (E.F.) and Tear on millimeter (1.8 mm) wave generation. In some ways the Compton papers are probably the most important work the REVIEW had published to that time. This is said without demeaning the work of Millikan on "e" and "h", but his work followed that of others, although in very superior and novel fashion. While H.A. Wilson had estimated "e" from the effect of electric field on C.T.R.'s fogs, it was Millikan's insight to go after individual drops (and Fletcher's substitution of oil). "h" and Einstein's equation had been pursued by several workers in photoelectricity with conflicting results; Millikan developed his vacuum scraper to make results definite and showing Einstein to be correct, as indeed had Richardson and Compton's brother (K.T.) shown earlier with ultraviolet light. A.H. seems to have arrived at his discovery pretty much full blown and essentially out of the blue.

There was apparently some anxiety in getting the work published expeditiously, for it was learned that someone in Europe was hard on his heels. Kevles indicates that pressure was put on the REVIEW Editor and his Board to make haste in getting it out, and that an eventual outcome, as well as concern over delay generally, was a new section in the journal: *Letters to the Editor*, showing up in the Index on the inside of the front cover (by then the familiar but dull green) for the first time in July of 1929. There was undoubtedly a consensus by contributors that there should be means for expediting the publication of work more important than the "usual." We of course today have an entire journal, PHYSICAL REVIEW *Letters*, devoted to such important and urgently needed contributions, descendent of the section in the REVIEW itself. In any event, Compton got there first with his publication and would rightly win his Nobel award for the discovery.

The first volume (21) of Fulcher's also had the Proceedings of three APS meetings in which the REVIEW figured. At the 117th meeting there was "extended discussion of methods of conducting the business of the Society with regard to offices of Secretary, Treasurer, and Managing Editor. But there is no clue on the gist of the talk. (Among the many elected to membership was one Lee A. DuBridge.) At the 118th meeting, in Cambridge, Gordon S. Fulcher was elected Managing Editor; no explanation given for Bedell's leaving the post. And at the 120th meeting, at the Bureau of Standards, APS dues went up—Fellows to $12 and Regulars to $9—because of the increased cost of *Science Abstracts*; since APS members account for a large part of the total subscription list they should assume their full share of support for the valuable abstracting service, at the same time providing additional funds for the PHYSICAL REVIEW . At the 117th meeting, an abstract indicates that, in this near modern era, vowels are still with us; G.W. Stewart discusses D.C. Miller's "Analysis of the Sustained Vowel e, as in Meet" by demonstrating "—the Variable Character of the vowel e," paper Number 59. There were significant contributed papers beyond those of Compton and of Nichols and Tear: Langmuir treats the effect of initial velocities in electron space charge, and a colleague, Kingdon, considers ions in the neutralization of space charge; Dushman, also colleague, derives thermodynamically the Richardson equation for thermionic emission, the constant "A" therein involving only universal constants; Kemble and Van Vleck worry about the specific heat of hydrogen vs temperature, while Tolman, Karrer, and Guernsey repeat in a different manner the electron inertia experiment of Tolman and Stewart on the electric current resulting in the rapid acceleration of a metal, to tighten certainty in the reality of the effect. K.T. Compton is working on the theory of the arc, and Millikan and his crew are now into experiments on slip, friction, and viscosity, rotating various surfaces in gases; that viscosity constant to its 3/2 power is still a worry to the man.

Under Fulcher in Volume 22, the journal loses its look of antiquity and quaintness, appearing very much as it did up through the Second World War. "Synopses" are now Abstracts; bold face type separates important sections in papers and, in the index, is used in paper titles; in the yearly index, authors (in bold face) are separated from subject index. The Contents for each issue appears as with Volume I, Number 1, Series I, on the front cover, a pale olive green in contrast to today's vivid green with the contents Table on the inside. The APS meeting abstracts in Proceedings begin to be published sensibly with an index of authors (and abstract number) following the abstracts, numbered serially, just as they are in today's *Bulletin of the American Physical Society* (publication resumed in 1925).

In Volume 22 Davisson is joined by Kunsman in scattering low speed electrons from platinum and magnesium in an apparatus, diagrammatic sketch of which reminds one of what is coming up later in Davisson's best known work, with Germer; the plots also remind one of that future paper. At Columbia it is still the Phoenix Physical Laboratories, where Terrill is studying the "Loss of Velocity of Cathode Rays in Matter." Van Vleck has "Two Notes on Quantum Conditions," Gregory Breit wonders "Are Quanta Unidirectional?," and Olmstead and K. T. Compton have a nice experiment on the "Radiation Potentials of Atomic Hydrogen," the molecular form dissociated in a hot oven; the excitation curves are not as pretty, nor as striking, as the usual Franck–Hertz plots but the authors find several excitation values, in complete and close agreement with Lyman's series and Bohr's picture. And Millikan considers the influence of molecular reflection at surfaces in "The Law of Fall of a Small Sphere through Gas"; the motivation is clear. In their many continuing and nontrivial contributions, the Comptons and Millikan were implicitly strong supporters of the REVIEW and lent it much credibility.

At the 123rd APS meeting in Chicago, the REVIEW in Volume 23 notes that Bohr was elected an honorary member of the Society, addressing the meeting on "The Quantum Theory of Atoms with Several Electrons." At the next meeting, Christmas in Cincinnati, the Proceedings record that John T. Tate was elected to the Board of Editors, beginning his long tour of duty for the journal. Compton (A. H.) and Hubbard in a regular contributed article are investigating the "Recoil of Electrons from Scattered X-rays," Millikan and Bowen (working down below 150 Å) are photographing "Extreme Ultra-violet Spectra," Leonard Loeb is measuring "The Mobility of Electrons in Helium," and J. C. Slater is theorizing on the "Compressibility of Alkali Halides," work done in Bridgman's laboratory as Slater's Ph.D. thesis. This is the first of many Slater papers in the REVIEW putting him on the path of his pioneering advances in quantum chemistry, with Pauling's contributions, pretty much an American development. Harvey Fletcher is now into acoustics at the to-be Bell Laboratories with "Physical Criterion for Determining the Pitch of a Musical Tone." By now, there are all manner of familiar names—at least to the older generation: Jauncey, Breit, Epstein, Stranathan, Eldridge, Page, and so on. We are practically at the modern era. The *Book Reviews* Department, which had appeared sporadically and diminishingly from "early on" in the REVIEW (as *New Books*) has become a regular feature, reviewing several books with each issue. Today, it has long since been gone, probably forever.

In Volume 24 (1924) Ehrenfest and Tolman write from Pasadena an article, "Weak Quantization," and Davisson and Germer give us "The Work Function of Oxide Coated (Ba and Sr) Platinum," from New York. In the next volume, Germer goes solo in an experimental study of the initial velocities of electrons emitted from hot bodies; the distribution is indeed Maxwellian. In the same vein, same volume, Hull and Williams, also into electrons from hot bodies, utilizing Schottky's schrott–shot–theory that there is noise in an unsaturated electron current because of the randomness and discreteness of the charges emitted, determine "e" to within a percent of Millikan's value—something of an accident but confirmation of the general idea and far more conclusive than a previous work in Germany by Hartmann. Kemble applies Bohr's correspondence principle to degenerate systems in a calculation of relative intensities of spectral lines in band spectra. Carl Eckart, disdaining corpuscles, has "The Wave Theory of the Compton Effect," referencing a similar notion (not yet *the* concept) of de Broglie, and a noted paper of Bohr, Krammers and Slater. Mulliken (u and e, not i and a) writes on "The Isotope Effect in Band Spectra," alongside another Millikan (i and a, not u and e) paper with Bowen. In the final volume of this pre-modern period of the REVIEW (Volume 26, 1925), J.B. Johnson also works on the shot effect, but at low frequencies where the flicker effect (positive ions) dominates (Hull and Williams were up at about a megahertz), which will lead him to his discovery of electrical noise in ordinary conductors a few years ahead. More and more familiar names: Hopfield, Birge, Becker, Blodgett, Hughes, Bowen, Slater, Allison, Foote, Mohler, Turner, Tuve, Van Vleck, Valesek, Richtmyer, the Zeleny's, Millikan (seemingly forever), the Comptons (also endlessly as with Millikan, much also to REVIEW advantage), and so on. With such a gallery, it is little wonder that the discipline of physics in America was finally on a firm footing, ready to take off, making the REVIEW's future secure. Another name, coming as only an abstract in this volume, is of interest: P. Debye (at MIT for a while) at the 131st APS meeting in New York has "Some Thoughts on the Diamagnetism of Gases at Low Pressures"; with World War II, he would be one of the European shining lights to move to this country, distinguishing Cornell's Chemistry Department by his presence.

Chapter 10
MINNESOTA YEARS
(1926–1945)

••

The PHYSICAL REVIEW in Volume 27, Second Series, carries the Proceedings of the winter 136th meeting, in Kansas City, of the American Physical Society and notes that John T. Tate, of the University of Minnesota, was elected for a three-year term to be Managing Editor of the PHYSICAL REVIEW. And it is on the cover above the names of the Board of Editors: John T. Tate, Managing Editor. He had not yet served out his three-year term on the Board. Nowhere is there a sign as to why Fulcher only served three years. His short tenure is a curious circumstance between the long terms of those on either side of him, Bedell and Tate. Nevertheless, in the Minutes of this 136th meeting, after "The Managing Editor of the PHYSICAL REVIEW presented the financial report of the year 1925 together with a report showing the progress of the REVIEW during the years 1923, 1924, and 1925,"

"On motion of Professor R. A. Millikan, the following resolution was unanimously passed:

Resolved, that the Physical Society express its appreciation to the retiring editor of the PHYSICAL REVIEW for the following specific reasons:

1. He has introduced an abstracting system which had proved a distinct contribution to bibliographic method and has made the PHYSICAL REVIEW in this respect a model already followed by many journals.
2. He has greatly improved the quality of papers appearing in the REVIEW, by an immense amount of detailed editorial work,

135

*and by insisting that contributors present their results with
brevity and elegance.*

*3. He has given a demonstration of value to other sciences in
that a journal, which a few years ago seemed unable to run without a subsidy, has been made to do so by attention to brevity and
other elements of quality. By these services Dr. Fulcher has made
an outstanding contribution to scientific progress.*

No such resolution or notice seems to have been made earlier in
regard to the stewardship of Frederick Bedell when he stepped
down. The apparent slight is somewhat distressing to one who knew
him and respected him, particularly when it appears that some of
the improvements Millikan credited to Fulcher had already been ini-
tiated under Bedell's management. Maybe as Board member, Fulcher
prodded the long-time Editor into them. Or was Millikan's resolution simply a gratuitous way of relieving Fulcher of his command? In
any event, that is what we have on the matter and John Tate
assumes the running of the journal. It was by now a mature publication; the contributions were solid and pertinent to the growth of
physics, not so much as an "industry" (there is that of course) as in
the understanding of the physical world. The REVIEW could well
take its place alongside the leading scientific journals, either here or
abroad.

Professor Tate was a middle-Westerner, graduating from the
University of Nebraska and earning a Master's degree there (the
Brace Laboratories of Physics—we've seen Brace) before going to
Europe for his Ph.D. degree at the University of Berlin. Following
two years as an Instructor and Assistant Professor at Nebraska, he
went to Minnesota as an Instructor, becoming Professor in five years,
a position he held when he assumed editorship of the REVIEW. From
1936 to 1939 he was Chairman of the Governing Board of the
American Institute of Physics, which he was instrumental in founding, and he was President of the Physical Society in 1939. He was
also Managing Editor of the *Reviews of Modern Physics* from its inception in July 1929 (with an issue given to Birge on the physical constants, A. H. Compton on corpuscular properties of light, and K. K.
Darrow on statistics in matter, radiation, and electricity). Tate's
research was directed largely at phenomena occurring in the impact
of electrons with atoms and molecules, the determination of appear-

ance potentials of various molecular ions, and the like, a number of his students having earned widely recognized reputations (e.g., A. O. Neir, Walter Brattain, H. D. Hagstrum). Elder members of the Society will recall the pleasure we took in his marriage as a widower of six years to Madeline Mitchell in 1945; she was the linchpin in running the Publications Office of the Institute and it seems everyone knew of her.

From this point on it is not clear how best to proceed with this REVIEW. Simply to present titles of a few papers from every volume would become boredom for the reader, physicist though he or she may be; it has perhaps already become that in what has already been culled of the early years. Not only that; it would be almost impossible. And yet major discoveries lie ahead in the pages of the REVIEW, major figures write, and the growth of American physics can be illustrated in a perusal of the journal volumes and citation of striking papers it has published, let us say up through the end of World War II, papers of such import that they must be recognized in any survey such as this, along with papers and notes of not such great moment but of interest nonetheless.

If there was any question about the world stature of the PHYSICAL REVIEW at the beginning of the Tate regime, there was surely none thereafter. During the period, quantum theory developed and came of age, nuclear physics and the physics of solids both matured, cosmic rays were a major activity, and modern particle physics was coming on. In all this, American physicists played an active and leading role, particularly the younger set, many of whom were still going to Europe for exposure to new ideas. The PHYSICAL REVIEW would chronicle much of the development, especially in the latter part of Tate's twenty-five years at the helm.

The discerning physicist scanning the papers summarized in *Science Abstracts* that came as part of the PHYSICAL REVIEW package with membership in the Physical Society could tell in 1925 and 1926 that great new ideas in physics were breaking in Europe. Our readers would find Abstract Number 1329 of 1925 interesting; it is that of Louis de Broglie's seminal suggestion from *Annales de Physique* (de Broglie has one PHYSICAL REVIEW contribution, in 1948 (Volume 76) on quantum electrodynamics, building on Feynman's relativistic cut-off in electrodynamics (Volume 74, 1948)). It is followed in short order by three of Schrödinger's great papers from *Annalen der Physik* as 1926 Abstracts, Number 1512, Number 2017, and Number 2024, the first being "Quantization as a Problem of Characteristic Values," the last being that in which he shows his approach and that of Heisenberg are really the same. Abstract Number 45 of 1926 is of the paper by Heisenberg from *Zeitschrift für Physik*, "On the Quantum Interpretation of Kinematical and Mechanical Relationship," developed further by Born and Jordan, and by Born, Heisenberg, and Jordan, Abstracts Number 447 and Number 1229, respectively. Then, still in 1926, came an Abstract (Number 1245) of a Dirac paper on Heisenberg's quantum mechanics and one of Pauli on the hydrogen spectrum by way of Heisenberg's mechanics, as Abstract Number 1330. Thus were the floodgates opened, developments to stir the interests of American physicists in, or heading for Europe, not to mention the eyes of REVIEW readers at home scanning their *Science Abstracts*.

Volume 27 (1926) of the PHYSICAL REVIEW, the first of Tate's many, may not have been as spectacular as these Abstracts of the year, nor even perhaps as the first volume of Fulcher's, but there were authors familiar to us in work of significance: D. M. Bose ("Kα Doublet Irregularity"); Linus Pauling (a two-page note on "The Dielectric Constant and Molecular Weight of Bromine Vapor"); Harold Urey ("The Structure of the Hydrogen Molecular Ion"— hydrogen, mass one that is; mass two was yet to come); Millikan and Eyring ("Laws Governing the Pulling of Electrons out of Metals by Intense Electric Fields"); Otis with Millikan, and Bowen also with Millikan ("High Frequency Rays of Cosmic Origin, I and II"—results of balloon flights and observations on Pike's Peak; Millikan was heading into his conflict with A. H. Compton, already into one with Hess and Kolhorster, Millikan believing in the "local origin of the radiation"); Bergen Davis (measuring the index of refraction for X-rays in an aluminum prism with a double crystal spectrometer, itself a big instrumental advance); and A. H. Taylor (again, and by now at the Naval Research Laboratory) with E. O. Hulbert (considering the propagation of radio waves over the earth).

The next volume (28, 1926) also had well known names but, in addition, incorporated more "meat"; in fact, the volume is loaded. There is a rather important but little known work of Carl Eckart, which was also done independently by Schrödinger, noted above. It is titled "Operator Calculus in Quantum Mechanics"; his treatment included not only the Born-Jordan matrix mechanics but also the "remarkable quantum conditions" of Schrödinger, showing them to be equivalent. In a later not too satisfactory paper on the hydrogen spectrum, he does not consider the "possibility that the electron itself may be in rotation" as suggested earlier in the year by Goudsmit and Uhlenbeck in a. communication sent to *Naturwissenschaften* in November 1925 by Ehrenfest and, four months later, to *Nature* by the discoverers. (A real lapse there in the work of the Nobel Committee.) Van Vleck treats the specific heat of hydrogen in the new quantum theory, and Paul Epstein considers the "Stark Effect from the Point of View of Schrödinger." Americans were not losing much time in joining the fray, unlike their participation in the period following Bohr's theory back in 1913. There is a nice paper by Schrödinger himself, from Zurich, on "An Undulatory Theory of the Mechanics of Atoms and Molecules" (his only contribution to the REVIEW). There is a paper on the band system of carbon monoxide by R. T. Birge, followed by an important one from E. U. Condon on a principle known to all molecular spectroscopists: "The Theory of the Intensity Distribution in Band Systems," an "outgrowth of a picture proposed by Franck...," stressing at the conclusion again that "my work is merely an extension of a leading thought on this subject by Professor J. Franck." This was followed by a "molecular" paper of R. Mulliken, "Electronic States and Band Spectrum Structure in Diatomic Molecules" (I and II); his many contributions to the REVIEW and elsewhere would lead him to his prize. The REVIEW Cumulative Index of 1950 will have as many entries of his as of his near namesake. By the latter, R. Millikan (we're having trouble keeping these vowels straight in here), there is a paper with more cosmic ray measurements; he had his prize. Still in this same volume, 28, E. O. Lawrence makes his first showing, from Yale on "The Ionization of Atoms by Electron Impact"; good enough, but not his greatest work. And Walter Schottky from Munich considers the flicker effect in J. B. Johnson's measurements. Another paper prescient in importance is that of Breit and Tuve: "A Test of the Evidence of the Conducting Layer," referring to the ionosphere and the propagation of radio waves around the earth. This was a pulse technique in which 1/1000 second pulses of radio waves were radiated upward, the time delay observed in the reception of the reflected signal to infer the layer(s) height—the technique antecedent of radar. It has been said that the radar concept originated with various others, including our A. H. Taylor of NRL. Indeed, Breit and Tuve acknowledge the help of Taylor who made "some of the circuit arrangements," but the concept would seem to belong to Breit and Tuve.

1927 is a big year, not only for the REVIEW but for physics in general. In Volume 29 quantum mechanics takes off. VanVleck considers the dielectric constant and magnetic susceptibilities in the new view; and Pauling, in the new way, treats the dielectric constant also but in the presence of a magnetic field. I. I. Rabi publishes his first paper, a nice piece, "On the Principle Magnetic Susceptibilities of Crystals" (thereby learning to always report one's work first at a meeting with an abstract) and, with Kronig, works out "The Symmetrical Top in Undulatory Mechanics." Henry Norris Russell, the great Princeton astronomer-spectroscopist (Russell–Saunders coupling) is in "On the Calculation of Spectroscopic Terms for Equivalent Electrons," appearing this year for the first time beyond his consultation with Compton on the latter's earth rotation–liquid inertia experiment we saw earlier. The fine structure in hydrogen, even after Sommerfeld, has an important role to play; W. V. Houston, much later to be at Rice Institute (he had a terrible time getting the folks in "Hewston" to pronounce his name "Hooston") but now in Pasadena, furthers it with a "Compound Interferometer for Fine Structure Work." R. C. Gibbs at Cornell will enter that picture (with Williams) but as for now writes with Harvey White on "Multiplets in Spectra." Gibbs' previous papers seem rather mundane, rather in the early Cornell mold, work to keep the journal filled. By now this was less of a problem, which did not help the financial state; at the Winter APS business meeting, covered in the REVIEW (Volume 29) the Editor reported a financial loss for the year, which would likely continue into future years, largely the result of the increase in number of papers published.

Volume 30 (1927) is another high point in the progress of our journal. After an initial announcement in *Nature* and an oral report at the Spring APS meeting in Washington, Davisson and Germer describe in the REVIEW their work in "The Diffraction of Electrons by a Crystal of Nickel." The patterns in the distribution of electrons reflected from an annealed crystal (the apparatus had broken, been repaired and baked for a long time to clean it up, converting the sample during the incidental anneal into a few large crystals) unmistakably showing selective reflection. The phenomenon had been predicted by Elsasser if de Broglie's hypothesis were correct; in fact, Elsasser believed that the plots of Davisson and Kunsman (cited from Volume 22) provided the evidence. Davisson could not agree with the view. Thompson (G. P.) in England, of course, also provided independent evidence in the pattern exhibited by a fine pencil of electrons passing through a thin film of gold. So waves were pretty clearly in, no doubt of it. Bateman wrote on the theoretical under-pinning of the new mechanics, getting de Broglie's and Schrödinger's wave equations from "A Modification of Gordon's Equations." Away from the quantum mechanics track, Houston uses his instru-mentation in a precision measurement of e/m from the H_α and the 4861A ionized helium spectrum lines; and F. L. Hunt, by now at the Bureau of Standards, measures other wavelengths—those of long wavelength X-rays, diffracted from a ruled grating, which A. H. Compton had shown could be done, thus paving the way to absolute wavelength measurements and accurate determination of crystal lattice spacings. In anti-climax, Davisson and Germer report on the thermionic work function of tungsten, correcting their work of Volume XX for the effective reduction of the barrier by the applied electric field (Schottky Effect). Costa, Smyth, and the Princeton Compton devised a "Mechanical Maxwell Demon"—two toothed wheels on a single rotating shaft to measure molecular velocities, a la Fizeau and the velocity of light. Concurrently and independently, John Eldridge was going them one better; later in the year he reports his method, employing several toothed discs along a rotating shaft, successively shifted in angle, a more effective velocity filter than the Princeton device, the results of which, in his view, had a precision "quite low and can hardly be said to constitute a verification of the law." And at Harvard, Chase repeats the Trouton–Noble relativity experiment on the basis of criticism of earlier work (including some of his own, Volume 28, made by Epstein, Volume 29). The experi-ment purports to check the "ether drift" by the torque exerted on a charged condenser suspended by a fine thread; the ether is not detected.

In spite of the previous discouraging financial note made by the Editor, the Proceedings of the 148th APS meeting, held in the Winter of 1927 at Nashville, tell in Volume 31 of his reporting a better year.

Based on membership growth, and notwithstanding the increased size of the journal, it should be financially okay until 1931; unfortunately, it was not to be, as the business meeting a year later would show. Interesting of the 148th is a resolution that was passed mandating that at each meeting there be a session devoted to Invited Papers, to prevent meetings from becoming too specialized; most papers were of great interest primarily to but a small number of attendees. One notes that among individuals voted in as members were J. R. Oppenheimer, E.C.G. Stueckelberg, and Jerold Zacharias. At the 149th meeting (February, New York) there was note made of the death of H. A. Lorentz, and among those made Fellowship members were Erwin Schrodinger and John C. Slater. This was a joint meeting with the Optical Society; as guests of the to-be Bell Laboratories, members were shown television (whirling discs) by our man Ives; a demonstration of talking motion pictures was also made. Shades of the future. (On a personal note: Volume 30 carries the Proceedings of the Physical Society meeting held, of all places, in Reno at the University of Nevada. Although a young resident of the place at the time, the writer recalls not a thing of it, much as his father must have been involved in the arrangements. But it came only a month after Lindbergh's famous flight, attendant hullabaloo over which may explain the diverted interest.)

More quantum mechanics figures in the next volume (31, 1928). J. R. Oppenheimer has three notes on the quantum theory of aperiodic effects, and one on the theory of electron capture. Condon does the physical pendulum in quantum terms and S. Goudsmit makes his first appearance in "Multiplet Separations for Equivalent Electrons in the Roentgen Doublet Law"; he's now at Michigan. Cosmic ray expeditions are in vogue; Millikan (again!) and Cameron report on their expedition to the Andes, looking at both altitude and latitude effects; the latter would come to be a source of embarrassment for Millikan. And there he is again with Eyring and Mackeown on more field emission in "Field Currents from Points." And A. T. Jones has something a bit out of the past: on the vibration of the bells of the Harkness Memorial Chime at Yale.

S. K. Allison in Volume 32 discusses X-ray line intensities in uranium; he would be somewhat differently concerned with the element fifteen years ahead. In the same volume, J. B. Johnson reveals "his" noise in "Thermal Agitation of Electricity in Conductors," reporting from what are known at last as the Bell Telephone Laboratories, in New York. Earlier (Volume 31) Joe Becker and Donald Mueller discussed from the same establishment "Electric Fields near Metallic Surfaces" in the matter of the work function. (Becker quite disagreed with Langmuir in this. In a 1935 Cornell symposium on electron emission at which the two disputed, one recalls Langmuir's criticizing Becker's "Alice in Wonderland picture" of the situation and still hear and see in the mind's eye, scrappy Becker coming forward, down from the rear, in rebuttal, crying "Oh, I've read the quantum mechanics, etc.", and launching into his argument.) Wadlund makes absolute X-ray wavelength measurements with the ruled grating, diffracting X-rays from a calcite crystal to get the grating space and thence obtaining the electron charge through knowledge of the Faraday. His value agrees with Millikan's, with slightly worse probable error. Condon in Volume 32 has more on Franck's principle which Condon had developed (Volume 28): "Nuclear Motions Associated with Electron Transitions in Diatomic Molecules." Lawrence and Beams measure a time lag of less than 3×10^{-9} seconds in the photoelectric effect. Wave mechanically, Breit treats the "Propagation of Schrödinger Waves in a Uniform Field of Force" and later treats Heisenberg in "The Principle of Uncertainty in Weyl's System," and Mulliken continues his molecular work with two papers, "Assignment of Quantum Numbers for Electrons in Molecules, I and II." More significant spectroscopy, however, is reported this year (1928) from the Indian subcontinent in the still young and obscure *Indian Journal of Physics*, conducted by C. V. Raman, M.S., D.Sc. (Hon.), F.R.S., previous contributor to the REVIEW. In the Indian journal, the conductor reports on "A New Radiation"—the effect now bearing his name. In the next year in this country, from Cal Tech, F. Rasetti writes significantly in "Incoherent Scattered Radiation from Diatomic Molecules" (Volume 34, 1929). Work with the phenomenon would become an important discipline in its own right.

Earlier in the year (1929, Volume 33), Gurney and Condon, from Princeton, had an important paper, independently of Gamow, on "Quantum Mechanics and Radioactive Disintegration"—particle tunneling. Davisson and Germer find no effect in "A Test for Polarization of Electron Waves in Reflection." Rojansky considers the Stark effect in hydrogen, and Kingdon and Charleton the

"Rectification of Radio Signals by a Vacuum Tube with Alkali Metal Vapor": oscillations of ions in the electron space charge minimum greatly increase the gain of the device for signals of relatively low frequency, an interesting effect, never very useful. M. L. Pool (Ohio) does a nice experiment on the lifetime of the metastable mercury atom in the presence of nitrogen, exciting the vapor with the total radiation from a mercury lamp and measuring the absorption of the 4047A line in the flash from a second lamp, a slotted wheel providing the time delay, the technique to become easy and commonplace in the days of lasers, a half century to come.

Uhlenbeck and Goudsmit in the second (Volume 34) of the year 1929 volumes take up a "Problem in Brownian Motion" studying in more detail the wandering deflection of a galvanometer mirror, treated first by Gerlach in 1927. And Goudsmit again, in the same volume, teams up with a new name—that of Robert F. Bacher—in a paper on "Separations in Hyperfine Structure." The two names are linked as authors of a later book of data important to atomic spectroscopists, *Atomic Energy States*. Also in spectroscopy, a significant paper by Slater, disdaining group theory in "The Theory of Complex Spectra." This was refereed by Condon (in the REVIEW offices at the time), which led directly to his interest in the field and to his great opus with Shortley, *The Theory of Atomic Spectra*, still a viable and valuable work. From Madras, India, a new name, not a frequent contributor to our journal but prolific and profound in other journals nonetheless, and to be recognized by a Swedish prize committee over fifty years later: S. Chandrasekhar with "A Generalized Form of the New Statistics." Also in Volume 34, Hughes and Rojansky treat the focusing of charged particles in a radial (cylindrical) electric field, obtaining the 127° focusing condition, a problem Purcell would extend to the spherical condenser a decade further along. In a combination of magnetic and electric focusing, K. T. Bainbridge was not successful in "A Search for Element 87 by Analysis of Positive Rays." A number of investigators had hunted unsuccessfully for "eka-cesium"; it would remain elusive even after discovery in 1940, as Francium. More familiar names: Brode, Harnwell, Edlefson, Korff, Lindsay, Pauling, Farnsworth, Lawrence.

With Volume 34, the REVIEW starts coming every two weeks; this in spite of some relief, instrumentation-wise, to be occasioned by the advent of a new American physics journal—the *Review of Scientific*

Instruments, Volume 1 coming in 1930, its first paper by Paul Klopsteg on the bifilar pendulum. Henceforth, there would be less instrumentation in the PHYSICAL REVIEW. And comprehensive survey papers in physics would be presented in another new journal, started the year before, as the *Reviews of Modern Physics*. At the winter APS meeting (New York, 1928, reported in Volume 33), besides re-electing Tate to another three-year term as REVIEW. Editor, a three-part resolution was passed: First, the Society authorized the publication of a supplement to the PHYSICAL REVIEW—it would "contain resumes of various fields, discussions of topics of unusual interest" and in a broad sense of "material that will give valuable aid to productive work in physics and yet that cannot appropriately be published in the PHYSICAL REVIEW"; Second, the Editor would be the Editor of the REVIEW working with a board; Third, subscription would be separate from that for the REVIEW. Thus, Tate became editor of an additional Physical Society journal; the supplement was called the *Reviews of Modern Physics*. Three *Reviews* now on the scene. Of interest is the title of an Invited Paper at these New York meetings: astronomer Walter S. Adams of Mt. Wilson gave an address on "A Large Telescope and its Possibilities"; 100 inches was not enough.

The inside of the back cover of the REVIEW December 15 issue gives subscription price for the PHYSICAL REVIEW supplement: $3.00/year to APS members, with the contents of the January issue to be "Electrical Discharges in Gases, I: Survey of Fundamental Processes," Compton and Langmuir (actually appeared in April); "Principles of Quantum Mechanics, Part II," Kemble and Hill (part I was already out); "Fundamentals of Band Spectrum Analysis," Mullikan; and "X-ray Analysis in Liquids," G. W. Stewart, our early contributor. The first volume of the supplement, subtitled *Reviews* of *Modern Physics*, held two issues, July and October, 1929. The first issue contained papers by Raymond Birge ("Probable Values of the General Physical Constants"), Arthur Compton ("The Corpuscular Properties of Light"), and Karl Darrow ("Statistical Theories of Matter, Radiation, and Electricity"). By the time of Volume 2, the journal had become the *Reviews of Modern Physics* (formerly the PHYSICAL REVIEW *Supplement*) and the planned, quarterly four issues duly appeared.

By 1930 a new Department had become an essential feature of the PHYSICAL REVIEW, which feature itself would ultimately evolve into another separate REVIEW supplement—PHYSICAL REVIEW *Letters*. The Department was simply *Letters to the Editor* and it was established in July 1929 in Volume 34 to facilitate and hasten the publication of work outstandingly important and urgently required. Of six letters in the initial "run," the first was "Wind Mixing and Diffusion in the Upper Atmosphere" by E. O. Hulbert, urgency of which perhaps somewhat compromised the intent of the Department.

At the head of the new department was this note: "Prompt publication of brief reports of important discoveries in physics may be secured by addressing them to this department. Closing dates for this department are, for the first issue of the month, the twenty-eighth of the preceding month; for the second issue, the thirteenth of the month. The Board of Editors does not hold itself responsible for the opinions expressed by the correspondents." Three days before date of publication! Hardly any outside refereeing. Today, the median delay time is about 180 days, the reports are not noted for brevity, we have rafts of referees, and the letters are not addressed to a department but are sent to the separate journal, PHYSICAL REVIEW *Letters*.

In the Minutes of the 1929 APS winter business meeting (Des Moines, REVIEW Volume 35, 1930) the Managing Editor reported that "the reception of the (PHYSICAL REVIEW) Supplement for six months was very gratifying." He reported increased speed of publication of letters and papers but at a loss of $2,400 to the REVIEW; there was an increase in the number of papers published and printing costs had escalated. Financial futures he was finding it harder to predict. (I am told that with the February 1930 issue of our journal, well before the AIP took it on, a page charge was instituted by the management—all of $2—in an effort to stave off or at least stem such deficits. In 1962 the page charge was up to $40; 1969—$60, this in spite of experiment started the previous year with cheaper, cold type, i.e., typewriter, composition and letter press printing; by 1982 it was $85; 1987–1993—$25, but no free reprints. This all was perhaps of no concern to the general APS member and so not brought to the Society for consideration or information at business meetings, to appear later as Minutes in the Proceedings, as would the unlike matter of dues, levied on every member. In any event, if the report is correct for February 1930, it was not a successful ploy.)

During the year 1930 (Volumes 35 and 36) Breit, Tuve, and Dahl reported, in a regular article, on a Tesla coil in oil under pressure with which high voltages of over 5×10^6 volts were obtainable. The first two authors, with Hafstad, follow the article with one on the application of the arrangement to vacuum tube use in possible ion acceleration. The equipment was not very stable so that Tuve later would report on further high voltage engineering with an accelerator in which step-up transformers were arranged in cascade, a much more satisfactory means to the end. C. C. Lauritson, with B. Cassen, capitalize on the former's earlier work with Bennett and report on a 600 kV X-ray tube and its radiation. Slater treats "Cohesion in Monovalent Metals" and Millikan finds cosmic ray intensity to be definitely independent of latitude and sideral time, ruling out production in the Milky Way or the Andromeda nebula. (Worrying about astronomy and cosmic rays, and well ahead in our chronology, Baade and Zwicky in an APS abstract, Volume 45, 1934, discuss supernovae and cosmic rays, proposing the concept of neutron stars.) Two quantum mechanics papers should be noted: one by Eckart again on the "Penetration of a Potential Barrier by Electrons"—field emission and the Fowler-Nordheim approach; and Philip Morse on "The Quantum Mechanics of Electrons in Crystals" with reference to work of Bloch, Bethe, Slater, Peierls, and Heisenberg. And Oppenheimer has an important paper in quantum electrodynamics: "Note on the Theory of the Interaction of Field and Matter." L. P. Smith at Cornell has discovered and elucidated "The Emission of Positive Ions from W and Mo," which would play a role the next year in the important development of a vacuum tube electrometer of exceedingly high input resistance, to replace the Comptons' electrostatic device.

Volume 36 (1931) carries the information concerning that tube. G. F. Metcalf and B. J. Thompson at G.E. have developed the famous FP-54 "Pliotron" and describe it in "A Low Grid Current Vacuum Tube." (The paper is immediately preceded by a paper authored by one John Bardeen, "On the Diffraction of Electromagnetic Circular Symmetric Waves by a Disc of Infinite Conductivity.") The Metcalf–Thompson paper is followed in the next volume (37) by DuBridge's use of the tube in "The Amplification of Small Currents"; he has gotten down to 5 x 10^{-18} amperes—30 electrons per second. The days of the Compton electrometer were numbered; soon they would be museum pieces, as indeed so has by today the FP-54. Besides this pioneering work of DuBridge, Volume 37 has other well known works. Zartman, at Berkeley, in a paper still frequently referred to, "A Direct Measurement of Molecular Velocities," obtains the Maxwellian distribution in the deposit of a beam of bismuth atoms falling on a glass plate mounted on the inside of a rapidly spinning drum opposite a slit cut in the drum wall diametrically opposite, through which the beam periodically passes as the slit crosses in front of the bismuth oven. Another experimental paper is that of Thomas Johnson in the same volume reporting from the Bartol Research Foundation on the "The Diffraction of Hydrogen Atoms" in which a beam of hydrogen atoms is diffracted by a lithium fluoride crystal; entire atoms show wave properties. The photographic record is presented, convincing enough. More striking photographs, however, are those of well-known Harvey White in his "Pictorial Representation of the Electron Cloud in Hydrogen-like Atoms," illustrating for "plumbers" the charge distribution in a way the mathematics does not. Lars Onsager, at Brown, has a divided paper, Parts I and II on "Reciprocal Relations in Irreversible Processes," an important contribution recognized in 1968 by the Nobel Committee for Chemistry. Slater and John Kirkwood write on "The Van der Waal's Forces in Gases" and Tate's aide, E. L. Hill, does a problem in the quantum mechanics of one-dimensional crystals. The volume also includes Slater's important paper on "Directed Valence in Polyatomic Molecules," paralleling Pauling's "Quantum Mechanics and the Chemical Bond," a title almost matching that of his well known, 1937, Baker Lectures given at Cornell: *The Nature of the Chemical Bond*. He and Slater were at the forefront setting the course of quantum chemistry.

More in Volume 37: LaPorte and Uhlenbeck make an "Application of Spinor Analysis to the Maxwell and Dirac Equations," which is followed by one of Podolsky on "A Tensor Form of Dirac's Equation." A Letter to the Editor from Einstein, Tolman, and Podolsky, all then at Pasadena, worries about some fundamental questions in "Knowledge of Past and Future in Quantum Mechanics," presaging a better known communication to come on the so-called EPR paradox. Tolman is worrying about other fundamental matters; he writes with Ehrenfest and Podolsky "On the Gravitational Field Produced by Light," and, abandoning similitude and dimensional homogeneity, considers the "Entropy of the Universe as a Whole," bringing us a fate that Millikan cannot face. Finding no latitude effect, the latter's cosmic rays are photons representing somehow "the birth of atoms," something akin to steady state: no end.

The Proceedings of the winter APS meeting (168th regular, 32nd annual, Cleveland, December 1930) were reported in the REVIEW as usual (Volume 37, still). The Managing Editor reported that the Society journals were being well received by the membership and by physicists in foreign countries. The financial situation had to be watched; the increased quantity of publication brought increasing financial deficits which had to be met by the generosity of individuals and organizations. The meeting voted favorably a significant, four-part resolution, approved and put forward by the Council:

1. *That the Society establish a Journal of Applied Physics to start July 1, 1931, edited by the Editor of the REVIEW.*
2. *Members could elect a subscription to the new journal or to the Reviews of Modern Physics in lieu of their membership subscription to the PHYSICAL REVIEW.*
3. *Approve in principle the formation of Sections in the Society by subjects, and encourage affiliation with local physics clubs, and*
4. *That the Council proposes the formation of an Institute of Physics for the purpose of coordinating various societies whose interests are primarily in the field of physics for the purpose of supporting their publications.*

An amendment added to (4) mandated that consideration be given to setting up the "Institute" under Section B of the A.A.A.S. Perhaps this was done and perhaps it was not; nothing came of the amendment.

At the APS Council meeting in New York, two months later, the REVIEW Proceedings note that the Chemical Foundation donated $1,425.42 to meet the deficit of the PHYSICAL REVIEW as of December 31, 1930, and the meeting moved "that the thanks of the council be expressed ... ." At the same meeting Einstein was elected an Honorary Member of the Society; in coincidence and in some contrast, the volume has note of the death of A. A. Michelson, a nice portrait accompanying Millikan's memorial statement. (Millikan also wrote a biographical Memoir for the N.A.S. series.) Volume 37 is large, of all that had been published, slightly exceeded only by the hefty 1800-some pages of Volume 36; 38 would come in at over 2300 pages.

In Volume 38 Heitler from Gottingen and Ohio State (he must have been spending a leave at Ohio, where L. H. Thomas and Alfred Lande held forth) has a paper on "Quantum Theory and Electron Pair Bond" and Slater continues bonding with another of his own, "Molecular Energy Levels and Valence Bonds." Oppenheimer writes a "Note on Light Quanta and the Electromagnetic Field"; Dirac's theory is cited; Oppenheimer had reviewed with high praise Dirac's book *The Principles of Quantum Mechanics* in the preceding REVIEW volume, " ... book largely an account of Dirac's own work ... account unitary and coherent ... in his papers alone one may find the key to the solution of any problem which the theory is competent to treat." (But not a book for beginners.) In this PHYSICAL REVIEW volume, Kaplan in "... Light of the Night Sky" identifies two bright auroral lines as arising from ionized oxygen, and R. W. Wood, still in there, makes notable improvements to Raman spectroscopic techniques, which were useful and valid up to the time of lasers. Tolman writes on the universe again, this time on a non-static, periodic model with reversible annihilation; perhaps to comfort Millikan? Two more, of rather greater importance, although non-theoretical: Sloan and Lawrence on the "Production of Heavy High Speed Ions Without the Use of High Voltages"—by linear accelerator after the notion of Wideroe; and then the landmark Letter to the Editor by Lawrence and Livingston— hydrogen ions of greater than half an MeV have been achieved in a magnetic field by the use of relatively low voltages. An interesting Letter is that of McMillan from Princeton on an electric field (F) configuration leading to a uniform deflecting force on a molecular beam, the molecules having Stark effect energy of aF^n; this would be tested with a beam of HCl molecules ($n = 2$) in a 1932 APS abstract. In the volume, 38, the Council meeting held at the APS gathering in Schenectady (Sept. 1931) is also covered. K. T. Compton reported for the American Institute of Physics, about to be launched; H. A. Barton from Cornell has been named first Director, John Tate to be Advisor on Publications. REVIEW Editor Tate reports on the new applied physics journal, *Physics*; it was "developing rapidly, with plenty of material for publication and about 900 subscriptions."

Physics, under Tate (with his Minnesota colleague, Buchta, as assistant editor), conducted by the American Physical Society, commenced publication in July 1931, three months previous to the Schenectady meeting and Tate's report. In the inaugural issue, after two "talk" papers—"Notes on Metallurgy and Physics" and "The Human Value of Physics" (A. H. Compton), the first technical paper was "Capillary Rise in Sands of Spherical Grains"; applied enough. But the label *Applied* did not appear as part of the journal's name until Volume 8, in 1937. With that volume, under the Editorship of Elmer Hutchisson, it became the new *Journal of Applied Physics*, taken over and published by the Institute (because, it is said, the "pure" physicists of the Physical Society wanted no part of such "practical" work—some stuffiness compared to the "early" days when such as the analysis of vowel sounds, tests of liquid air plants, and evaluation of building brick made quite welcome contributions to the PHYSICAL REVIEW). Hutchisson opened with an editorial, "let us go forward," leaning on the last paragraph of Henry Rowland's first presidential address to the Society quoted earlier, which paragraph ("Let us go forward ... , etc."), in a box, the editorial surrounded. The first technical article then followed, one on high speed photography by a trio of MIT men: Edgerton, Germeshausen, and Grier, now probably better known as proprietors of a "high tech" firm passing under the name of E.G. & G.

Under the Directorship of Henry Barton, come from a Cornell professorship, the American Institute of Physics got underway in 1932 and took over management of the by now growing number of American physics publications besides our REVIEW: the *Reviews of Modern Physics* (1929), also under Tate, the *Review of Scientific Instruments* (1930) under Paul Foote (and F. K. Richtmyer three years later), the *American Physics Teacher* (1933), the *Journal of Chemical Physics* (1933), and sundry others founded in following years. On the inside back cover of the PHYSICAL REVIEW for April 15, 1932, a little after the fact, is "AN IMPORTANT NOTICE TO AUTHORS AND SUBSCRIBERS," notifying them that "After April 1, 1932, the American Institute of Physics will take over the editorial mechanics and business management of the PHYSICAL REVIEW," offering addresses to which future correspondence should be sent. A few issues later there is a back cover notice that the Institute has taken over the "editoral mechanics" and "business management." But it wasn't until the August 1st issue that the front cover reflected the change; no longer was it, as had been proclaimed until now the PHYSICAL REVIEW , "Conducted by the American Physical Society," it was now "Published by the American Institute of Physics, Incorporated, under the Editorial Supervision of the American Physical Society," changed within a year to "Published for the American Physical Society by the American Institute of Physics, Incorporated." It was so noted with each issue then until July of 1980, when it was, as it has been since, "Published for the American Physical Society through the American Institute of Physics." So also was it with the *Reviews of Modern Physics*. The slight change in wording seems the resolution of some disagreement over, and presumably puts right the question of, who was who in the matter of publishing the journals, and now states correctly the situation as had existed since the fifties when Goudsmit became Editor. We had come a long way from its "Conduction" by the trio of its first editors, under the auspices of Cornell University.

The taking over of the *Journal of Physical Chemistry* by the AIP is not devoid of interest or of connection to our REVIEW story. John Servos, in his article on Bancroft and the *Journal*, cited in our Introduction, tells of some of the contention associated with the founding of the magazine and the end of Bancroft's outdated editorship of his journal. There was need for more adequate and modern coverage of the domain overlapping physics and chemistry; the Institute proposed that it take over Bancroft's journal, rescuing it from almost certain severe financial difficulty and giving the Institute its forum in the area. Bancroft would have nothing of it; his *Journal* was for chemists and not physicists. Rather, he argued, the PHYSICAL REVIEW should install a section given to chemical physics, papers somewhat mathematical in nature. In due course, the section might then evolve into a separate journal. For a short period this seemed the wise course on which to proceed. Fortunately, perhaps, for the PHYSICAL REVIEW, the Institute did not stay the course; a few months later Barton went ahead with plans for a journal to include appropriate physical-chemical papers which might otherwise appear in either the REVIEW, the *Journal of the American Chemical Society*, or the *Journal of Physical Chemistry*. Reaction was mixed; Birge, for instance, did not like it, worried that specialization would lead to the destruction of science, while Urey, who would become editor of the new journal, is quoted as saying that to publish a physical paper in the Bancroft journal was "burial without a tombstone." Servos' story of the subsequent maneuvering makes for interesting reading, but it is beyond further recital here as regards our PHYSICAL REVIEW. The upshot was that the REVIEW did not come to have a "chemical" section, the *Journal of Chemical Physics* was born, and Bancroft capitulated, giving over to the American Chemical Society his journal, which came to flourish under the new editorship of S. C. Lind, radio-chemist in the Bureau of Mines.

1932 was a great year in physics; Kevles calls it the Miraculous Year. In the U.S. we now had 2500 physicists—three times what the number was at the beginning of the 20's decade—and our REVIEW was no longer a "back water, if not joke, publication"; European centers eagerly looked forward to its biweekly arrival (citing Rhodes in his "Making of the Atomic Bomb"). For our journal, the issue of January 1 (Volume 39; in this year there was a new volume of the REVIEW every three months. But the quarterly volume procedure did not take; a year later we were back to a new volume every six months) has a Letter to the Editor from Urey, Brickwidde, and Murphy; hydrogen of mass two has been discovered. In the next volume, 40, they write their article "Hydrogen Isotope of Mass Two and its Concentration." In the next article but one of that volume, there is Lawrence and Livingston with "Production of High Speed Light Ions Without the Use of High Voltages"; they wax euphoric, seeing the way to protons of 25 MeV in a machine using a 14 kG magnetic field, 50 kV across the Dees at 14 meters wavelength. Actually, not quite so simple. The next article but two is by prolific Robert Mulliken again on the "Electronic Structure of Polyatomic Molecules and Valence." The next article but one is by Blodgett and Langmuir on the "Accommodation Coefficient of Hydrogen; a Sensitive Detector of Surface Films." Four future Nobelists in one *issue*; Langmuir would be awarded his later in this great year. Another shows up later in the *volume* when Anderson with Millikan (who'd had his for nine years already) write on "Cosmic Ray Energies and Their Bearing on Photon and Neutron Hypotheses."

Across the way, in England, Chadwick had but recently reported in *Nature* his discovery of the neutron, "Radiation" strangely thought by Mme. Curie-Joliet and husband, and reported by them a bit earlier, to be in some way high energy gamma rays. Great year indeed! Another future Nobelist (Chemistry) shows up from Darmstadt in Gerhard Herzberg's Letter on the "Rotational Structure and Pre-dissociation of the P_2 Molecule." He would wind up in Canada, his work much akin to that of Mulliken, and receive his prize two years after Mulliken's, also in Chemistry. Besides the papers cited, there were others in this notable volume. G. F. Metcalf had an interesting one on increasing the charge sensitivity of his FP-54 electrometer vacuum tube through the use of what we now call feedback—here positive in nature. Part of the output of a second stage of amplification is fed back to the input to aid in charging up the input capacity. A patent was granted in 1922 to Brillouin on the idea of this capacitance neutralization, but the idea was not well known. Metcalf saw no obstacle to achieving a sensitivity of one electron per mm scale deflection of his galvanometer, but there is no report of such sensitivity having been actually reached. Campbell and Houston have an article on a precision determination of e/m by way of the Zeeman effect, and A. H. Compton is getting into the cosmic ray act, contributing a paper (with Bennett and Stearns) on cosmic ray ionization as a function of temperature and pressure.

In PHYSICAL REVIEW Volume 40 Birge gives an update on the most probable values of e, h, e/m, and α. And in this volume there was another new Department instituted in the journal, announced on the front cover of the first issue: "A New Service to the Readers": *Current Literature in Physics*, in which were reproduced, as in the early *Journal of Physical Chemistry*, index pages of a few of the better known journals. A notable entry: in our Volume 40, for the *Nature* index of February 27, there it is—J. Chadwick: "Possible Existence of a Neutron." But the Department, like the new quarterly volumes, did not last; although it seems a good idea, it was abandoned a few years later. *Book Reviews* were still with us, at the time a fairly large Department, soon also to die; in a single six-month volume, up to 50 or 60 (66 in Volume 38) books might be reviewed, many viable even today. We have in the several hundred reviewed (listed under Book Reviews in the Cumulative Index 1921–1950) books such as Dirac's *Principles* we spoke of, Van Vleck's *Theory of Electric and Magnetic*

Susceptibilities, Bateman's *Partial Differential Equations of Mathematical Physics*, Brillouin's *Les Statistiques Quantiques* and Hughes and DuBridge's *Photoelectric Phenomena*. Less memorable: Mary Crehore Bedell's *Modern Gypsies* and Person's *General Physics for Home Economics Students*. One REVIEW in the Department near the end of its life, of more than usual interest and the first in the year 1932 in Volume 39: George Gamow's *Constitution of Atomic Nucleii* (Oxford Press), not quite viable fifty years later. It was reviewed by Van Vleck, who complains about the complete lack of any reference to theoretical work; experimental work is well referenced. And nowhere, the reviewer states, is mention made of Gurney and Condon. One paragraph of the review is particularly catching; electrons don't behave well: "There is even considerable evidence pointing to the conclusion that in the nucleus the electrons lose completely all their normal properties. In fact, the law of conservation of energy fails unless one postulates the Pauli neutron, which incidentally is a hypothesis too recent for mention in Gamow's book. In the original manuscript, all paragraphs dealing with nuclear electrons were set off by a skull and cross bones"—(sounds like Gamow; he used the device elsewhere with suspect ideas he presented)—"but in the actual printing this has, alas, been changed to the rather colorless symbol ~." Pauli's particle is of course the neutrino and Chadwick's particle is even too recent for inclusion in Van Vleck's review.

In Volume 41 of the REVIEW, 1932, Anderson writes on "Energies of Cosmic Ray Particles," worrying about the interpretation of some of his cloud chamber pictures; there are positive particles, which must be protons. He is on the track (in more ways than one). And Compton heats up his controversy with Millikan with a Letter reporting progress in his own survey of cosmic rays; he has teams all over the world and disagrees with Millikan on the intensity *versus* altitude, and more importantly, on the presence of a latitude effect. For him, cosmic rays must be charged particles and not photons, for the less penetrating rays are the more strongly affected, presumably by the earth's magnetic field. J. L. Dunham, at Harvard and acknowledging Kemble's help, considers the Wentzel, Kramers, and Brillouin method, our WKB means, for solving the wave equation, extending it to higher approximations, and he applies it to a calculation of the energy levels of the rotating vibrator in a paper which follows it. Darol Froman from McGill observes "The Faraday

Rotation with X-rays," confirming evidence in earlier work suggesting such a rotation. Dennison and Uhlenbeck write on "The Two Minimum Problem and the Ammonia Molecule," a matter on which Townes would capitalize fifteen or twenty years down the line. And in a Letter to the Editor from Cal Tech, Bacher and Condon decide in somewhat convoluted argument that the most probable value for the spin of the neutron is 1/2, deciding that that of the electron in the nucleus (in combination with a proton to form the neutron) must be zero.

With the last of the quarterly volumes of 1932 (Volume 42), the fortunes of the PHYSICAL REVIEW will change again; publication of the journal is assumed by the newly established American Institute of Physics, which was taking on other journals, as noted some pages back. Little change to the REVIEW would be apparent, however, until the beginning of the new year. In this last volume of 1932, Bainbridge leads off in a measurement of the isotopic mass of the new hydrogen, and Kennedy and Thorndike follow with their "Experimental Establishment of the Relativity of Time," another interferometer set up; arms of quite unequal length. Lawrence and Livingston in a Letter with Milton White, have disintegrated lithium with their high speed protons, confirming the results of Cockroft and Walton; and in another Letter, Child, still at it, discusses the luminosity of the sodium flame.

By now, save for untiring Millikan, whose persistence and versatility surprises, the old guard is fading away, perhaps still with us but no longer at work. R. W. Wood is still heard from once in a while—he's pretty versatile also; in 1939 he will have a lecture demonstration on the resonance radiation of sodium vapor. But for the most part, the new physics has passed them all by. For us today it is a far more interesting and exciting discipline which we are now here sketching than that in which they indulged and worked. Yet we should not belittle their endeavors; we are the benefactors and the PHYSICAL REVIEW exists by virtue of their contributions, feeble though some of them may seem to be. "The old order changeth, yielding place to new"; somewhat sad but true. Enough of such sentiment.

It is in 1933 with Volume 43, with the REVIEW now under AIP auspices, when it is clear that change has come to the journal. It now comes again as two volumes per year, pages are larger and the articles are printed in two columns per page width rather than in full page width without a break. And we appear in the bright green cover (Index on the rear) which makes it easy to spot in Library journal racks. Advertising has disappeared, the July 1 issue telling on the inside front cover that advertising is not accepted. This, in spite of the fact that nine months earlier the journal, back in its advertising pages, carried a request urging that readers encourage advertisers in the use of REVIEW pages, suggesting four possible approaches. But with the AIP in the act from here on, the *Review of Scientific Instruments* (and the *Journal of Applied Physics*, 1937 to 1970) would be carrying the advertising, in major support of that journal. Disappearing from the PHYSICAL REVIEW also were *Book Reviews* and *Current Literature in Physics*, the latter not yet a feature for even two years. These last were rather nice Departments to have but apparently it was felt they were not worth the added expense. In the Minutes of the Atlantic City, end of 1932 meeting reported in the volume, the REVIEW Managing Editor reported on conditions of the journals of the Society, telling of a deficit of $3,000 as of the previous May 1; it had increased by about $750 per month since that time. Speaking of the journal's format, he said that the double column page would save money over full printed width of the page; it made for an easier setup in printing and it was easier to read as well. So costs should go down but "not enough to enable the PHYSICAL REVIEW to continue putting out the same number of pages without incurring a large deficit." The Minutes for the earlier meeting in Chicago (Volume 42) mention Tate's reviewing the returns on "The publication questionnaire," his pointing out the probable savings to be secured with the new program under the American Institute of Physics. The establishment of an Institute had apparently been made, for the Minutes of the New Orleans meeting, still more previous (December 1931, Volume 39), noted that the four resolutions voted the previous year, and seen above, had been carried out, and as an aside, he noted that the REVIEW, the *Reviews of Modern Physics* and *Physics* were "being well received with considerable increase of published material" and the over $6,000 deficit (maybe Fulcher should have still been on the job) "was largely covered through the generosity of the Chemical Foundation" (again). The page charge allegedly installed two years earlier was not doing all that had been hoped. At any rate, from the publications questionnaire circulated by Tate to the membership, he said that two-thirds of the 1050 respondents indicated they'd rather drop receiving the REVIEW than to have dues go up by $2 in order to continue receiving it. Seems a strange response; why else be in the Society? Anyway, the Council decided it would be unwise to raise dues at that point. This left the editor with but one recourse he said, namely, reduce the number of pages, and he thought that could be done without impairing publications. It is not quite clear how dropping the advertising saved money but its disappearance did lead to an uncluttered journal appearance. The REVIEW now had very much our today's look about it.

Under the new auspices, the REVIEW in Volume 43 has a paper from MIT by Van de Graff, Compton, and Van Atta on "The Electrostatic Production of High Voltages for Nuclear Investigations"; a 10 MeV generator is under construction at Round Hill (today part of that installation is in the Boston Museum of Science); Hiram Edwards (Berkeley) reports in a Letter on aluminum-magnesium mirrors produced by evaporation (no references given, but Cornell and Johns Hopkins argue over whether the former's Robley Williams or the latter's John Strong originated the procedure—suffice it to say that evaporation is the universal method for coating astronomical mirrors these days); E. Wigner and F. Seitz write "On the Constitution of Metallic Sodium"; Dunnington describes "A New Deflection Method for Determination of e/m of an Electron," a precision method based on the cyclotron principle of constant time of flight, independent of ion energy, around a given angular deflection in a magnetic field; it utilizes an RF voltage, the frequency of which, along with knowledge of the geometry and magnetic field, enters into the determination and precision. Tolman and H. P. Robertson have "On the Interpretation of Heat in Relativistic Thermodynamics," tending toward their cosmological interests; from the USSR, Frenkel has a possible explanation for superconductivity; Compton, immediately preceding Dunnington, reports finally on his geographical survey of the cosmic radiation, fairly strongly showing a latitude effect (in the survey, two men were killed at their station on Mt. McKinley by falling into a crevasse); S. Ballard and Harvey White have a nice series of pictures in a vacuum ultra-violet study on the "Isotope Effect in the Lyman Series," six members clearly split in a source having a mixture of the two isotopes. In a Letter, Tuve urges a search for Dirac's "isolated magnetic pole" (which Dirac would write more on (from Princeton in 1946) in Volume 74 in his "Theory of Single Poles"), and in another, much more significant Letter-length article, abstract by the Editor, Anderson has found a vacancy in Dirac's sea of negative energy electrons; it is to everyone's surprise, including Dirac's. A brief announcement had been made six months earlier in *Science*. In a famous cloud chamber photo reproduced in the REVIEW communication, a particle track cuts through a lead plate, curving in a magnetic field the more on one side of the plate than the other, thus giving the direction of the particle flight, it is positive and the ionization is too low for it to be a proton. The positron is found; the Letter-length article details: "The Positive Electron," labeled the positron, research suggested by Professor R. A. Millikan. It quickly entered theory; in Volume 44 Fermi and Uhlenbeck have a Letter, " On the Recombination of Electrons and Positrons." In Volume 43, another important Letter, this from a future Chemistry Nobel Prize winner: Giauque, who with MacDougall reports on attainment of low temperatures by adiabatic demagnetization; using gadolinium sulphate, they get down to 0.25°K.

With heavy hydrogen becoming available (in small quantities), the fine structure of the Balmer spectral lines is being pursued. Spedding, Shane, and Grace report on the Balmer alpha line for the mass two isotope in a Letter of Volume 44, and in the next volume, Houston and Hsieh at Cal Tech, and Robley Williams with R. C. Gibbs at Cornell give their reports on the fine structure of the lines. From the analysis of the structure for $H^1\alpha$ and $H^2\alpha$, the spectroscopists infer that the peaks observed are not quite where theory says they should be, even though the theories of both Dirac and Sommerfeld amazingly agree on the component splittings, in spite of their divergent theoretical concepts. The peaks' separation appears to be about 6% low, a conclusion re-enforced the next year (1935, Volume 48) in another Williams–Gibbs Letter (and an extended analysis by Williams in Volume 54) showing a clearly resolved bump on the side of one of the main peaks seen with a liquid air cooled discharge, allowing a considerably better determination to be made; the discrepancy is only slightly less than earlier suggested. In the Houston-Hsieh paper (Volume 45) they report the prescient suggestion of Bohr and Oppenheimer that perhaps the interaction of the electron with its own radiation should somehow be brought into the theory, speculating that a shift in the relative displacement of the lines would be about α times the fine structure separation. Not quite right, but still ... . Four years later, S. Pasternack suggests in a Letter (Volume 54, as with Williams' final story) that if the $2S^{1/2}$ level is displaced about 0.03 cm^{-1} from Dirac's position, experiment and theory could be brought together. But it would be almost twenty years (Volume 72, 1947) before the experiment demonstrating the truth of the suggestion is stunningly performed by Lamb and Retherford, to set off another revolution in fundamental physics. (In that same 1947 volume, Kusch and Foley contribute to the revolution by suggesting that g for the electron may not be exactly two.) Concurrent with these papers were other developments. Slater extends the work of Wigner and Seitz in "Electronic Energy Bands in Metals" (Volume 45, 1934). C. E. Cleeton and N. H. Williams, utilizing a split plate magnetron, observe the absorption spectrum of ammonia at 1.1 cm wavelength, as antecedent to the work of Townes, also twenty years ahead. Lawrence and Livingston report on their 27" cyclotron (the machine described later in the volume) in a Letter preceding the first of Williams–Gibbs; the Berkeley pair bombard various targets with their 3 MeV "deutons," resulting in the consequent emission of protons and neutrons; in a Letter with G. N. Lewis and Henderson they disintegrate "deutons" with protons, the four elaborating in a paper of the next issue. The cyclotron is coming into its own. Sandwiched in between this last paper and that of Houston–Hsieh above, is the important one of Oppenheimer and W. Furry, "On the Theory of the Electron and the Positive."

In Boston at the winter 1932 APS meeting (189th regular, 35th annual) the Managing Editor reported in Volume 45 that it was no longer his duty to report on the financial states of the various publications of the Society; it was now in the hands of the Institute. But he pointed out that in the first half of the year the average length of papers was 6.93 pages; in the last half it was down to 6.19 pages. There were advantages and disadvantages in this which he summarized: "Among other things it resulted in an accumulation of unprinted papers in the hands of authors for correction."

Rabi, Kellogg, and Zacharias (Volume 45, 1934) in a single inhomogeneous magnetic deflecting field measure the magnetic moment of the proton and the "deuton" (it will be a while yet before we know the nucleus of hydrogen mass two as the deuteron). It won't be long before the Columbia group greatly improves on the method and the precision. Also in here one finds a rather important paper in ion optics by Smythe (W. R.), Rumbaugh, and West, from Cal Tech. They have developed a "High Intensity Mass Spectrometer" which employs a new magnetic means of focusing ions from an extended region onto a slit; and for the electron beam creating the ions, an electron source was developed using a rubber membrane analogue for the electric fields (sans space charge) resulting in a gun rather much like that J. R. Pierce devised with electrolytic tank in the late thirties for space charge limited flow. The Cal Tech trio separated potassium 39, obtaining beams on the exit slit of up to 0.1 ma, allowing for nuclear disintegration work on the separated isotope. (Smythe with Mattuch had earlier built an all-electric mass spectrograph (Volume 40) using RF deflecting fields as a velocity filter, a means discussed by Smythe back in Volume 28 (1926).) In Volume 45, James Franck and R. W. Wood observe a difference in the long wavelength limit of the UV absorption of heavy and light water vapor, interpreted by a loose application of the Franck-Condon principle. O'Bryan and Skinner's "Characteristic X-rays from Metals in the Extreme Ultraviolet" is reported, followed by Jones, Mott, and Skinner with "A Theory of the Form of the X-ray Emission Bands of Metals." (Mott would be awarded a Nobel Prize in 1977 for his contributions to solid state physics.) In the same volume, Tuve, Hafstad and Dahl feel they have confirmation for the existence of a stable, still heavier hydrogen isotope, H^3. At this point it is largely speculation and wrong; the isotope would become known as tritium and turn out to be radioactive; it will be straightened out in another ten years. Their opinion was based on range-energy measurements of hydrogen ions accelerated in their cascade generator. In 1937, Rutherford will ask Aston to look for it; he will find nothing. There had been an earlier report of its discovery in Letters by Latimer and Young (Volume 44) and by Smith, Bleakney, and Lozier, (Volume 45), who reported it from a mass spectrograph at perhaps a few parts per billion in unseparated hydrogen. The Latimer-Young report employed the questionable magneto-optic effect of Alabama's Fred Allison, who claimed detection and discovery of element 87—"eka-cesium"—in the technique. But the effect was ultimately shown to be non-existent in some work of D. Morey under Merritt at Cornell. In 1940 the element would be discovered in France, and named Francium naturally, by Marguerite Perey in the disintegration of actinium; it is β-active with a half life of 21 minutes. The natural abundance in the earth's crust is thereby pretty small; perhaps a few tens of grams *in toto*. Nevertheless, in the search for H^3 we begin to see inklings of things to come in nuclear physics—some pretty frightening. There is a Letter from Chicago (Volume 46, 1934) by A. V. Grosse and M. Agruss on "The Chemistry of Element 93" and Fermi's discovery of it in the bombardment of uranium with neutrons. One is reminded of Fermi, in company of a friend, seeing for the first time a bust of himself put up somewhere in Rome; "There's the man who never discovered fission," he remarked. Perhaps a fortunate circumstance; given the state of the world as it degenerated in the last half of the decade, it would have been far too early to have had that knowledge in hand. In other areas of physics there are articles by H. P. Robertson from Princeton ("An Indeterminacy Relation for Several Observables and Its Classical Interpretation"), Wigner and Seitz, also Princeton, with more on metallic sodium, and a Letter from L. Marton, Brussels, reporting on first work with an electron microscope on biological objects; magnification with resolution considerably shy of what we can achieve these days. Five years later he would be at the RCA Laboratories and telling of a new instrument reaching 50 Å resolution. There was other work of biological impact: Lawrence reports in another Letter of the Volume on production (by deuton bombardment) of radioactive sodium; a nice half-life and 5 MeV gamma ray emission, both convenient for radiation therapy. The work is detailed further in Volume 47. The medical implications did no harm in funds-hunting for cyclotron construction.

Letters to the Editor was a too popular feature of our journal. Facing Marton's Letter in the Section was a full page admonishment "To Contributors" from the Editors. Five years of the Section had seen nice growth, but now, concerns: the most serious: "... growing tendency among contributors to be satisfied with the hasty, incomplete, and often inadequate record of their investigations ..." which Letters provides. Few enjoy writing up the record and the report "... when the primary urge to secure priority can be satisfied by dashing off a Letter to the Editor." Further, "It was intended neither that it (Letters) be a place for the preliminary announcement of *all* work, nor that, in the fields covered, it should replace more formal and critical articles. If the present tendency to record much of the important work in several laboratories by a series of Letters to the Editor of gradually increasing length is continued ..." standards of the PHYSICAL REVIEW will be "seriously lowered." The Editors ask for cooperation in achieving the aims of the Section.

Early in 1935, Hans Bethe arrived in this country, setting up shop at Cornell, and it was not long before he commenced writing the innumerable papers which appear under his name in REVIEW pages. In Volume 47 there is already a first Letter, on the "Masses of Light Atoms from Transmutation Data," and a long paper, "Theory of Disintegration of Nucleii by Neutrons." (At a Cornell pre-election party (Landon *versus* Roosevelt *versus* Browder *versus* Thomas), Bethe was spokesman for the Communist, on the basis of his "New Masses.") Also in Volume 47 is the famous, still controversial paper (48 related papers in 1985): that of Einstein, Podolsky, and Rosen on their (EPR) paradox—"Can the Quantum Mechanical Description of Physical Reality be Considered Complete?" A Letter only two issues later by Kemble says that it is complete, and another by Ruark in the next volume says it is either a matter of personal choice or definition. From Copenhagen, Bohr also in Volume 48, comments in an article examining the view from the standpoint of his complimentarity; he thinks the quantum mechanical description fills all "Rational demands of completeness"; their (EPR) criterion of physical reality contains an "essential ambiguity when applied to quantum phenomena." But it is still being argued over and experiments recently attempt illuminating and resolving the paradox.

During the year, the value of "e" is still, after all these years, drawing attention: In Volume 47, Birge and McMillan have a Letter on Schopper's new determination of the charge by the old means of getting the total charge carried by a known number of α-particles) and come up with a best value for the constant, almost 1% off from the X-ray value. Ruark, four pages earlier, worries in his Letter about x-ray wavelengths used in the latter determination, but Bearden, in a later Letter, refracting X-rays from a diamond prism and diffracting them from a ruled grating, only strengthens the X-ray wavelength scale and comes to new values of N_0, e, and h. His paper, which follows in Volume 48, details this work with large ruled gratings (in which R. W. Wood has had a hand) and gives us $e = 4.805 \times 10^{-10}$ esu, in contrast to the oil drop value of 4.768×10^{-10} esu, which Birge and McMillan had decided on in Volume 47. (Another Letter in that volume of importance is again by Giauque and MacDougall on establishing a true thermodynamic temperature scale in the region below 1°K.) In yet another Letter in Volume 48, Birge draws attention to Kellstrom's re-determination of the viscosity of air; the oil drop value comes up in close agreement with Bearden's value, which, except for the coming *coup de grace* in Bearden's own determination of the viscosity, agreeing with Kellstrom, about closes the nagging problem. Other Letters in Volume 48: Fermi, from Michigan, on the "Recombination of Neutron and Proton," a Columbia group (listed below, here also including Segre), with the familiar rotating paired toothed wheels arrangement, measure some slow neutron velocities; Zwicky wonders where negative protons can be found; and Beth has the "Direct Determination of the Angular Momentum of Light"—a torsion balance carrying a doubly refracting medium twists when the medium changes the state of polarization of an incident circularly polarized beam. The Columbia team of Dunning, Pegram, Fink, and Mitchell, report in a long paper on the "Interaction of Neutrons with Matter"; Einstein and Rosen discuss the "Particle Problem in General Relativity"; and Tuve and company write on their high voltage installation and their results on carbon radioactivity and other transmutations induced by accelerated protons. Konopinski and Uhlenbeck have "On the Fermi Theory of β-radioactivity (I)"; part II would come quite later, in 1941 (Volume 59). There is coming to be more and more nuclear physics. Quantum electrodynamics is worrying; in adjoining papers in Volume 48, Serber and Uehling consider modifications to Maxwell's equations, and polarization effects arising from Dirac's positron theory.

At the December, 1934, Pittsburgh (Volume 47) APS meeting there was no financial report from the REVIEW Managing Editor, as he had announced the previous year would be the case, but he took the opportunity to express his appreciation of the cooperation of the Society members in condensing their papers and writing them concisely so that it was possible to publish a considerably larger number of papers in the same number of pages than in the year before. A year later at the next business meeting, in St. Louis, he reported that the cost of putting out the three Society publications would be less than $5 per member. The low cost came from two circumstances: they added the cost of postage to foreign subscribers, and the amount published was just about what it had been before; he indicated that the fortunate state of affairs would not last.

In 1936, R. P. Johnson and William Shockley (Volume 49) report on a simple cylindrical, electron microscope for observing electron emission patterns from an axial filament, Breit and Wigner capture slow neutrons, Gamow and Teller in their article come up with a new β-decay mode (G–T) and selection rules for such disintegrations, Serber notes "Positron Theory and Proper Energies," Pauli and Rose have simply "The Positron Theory" and John Bardeen has the "Theory of the Work Function: The Surface Double Layer." Nothing bad about the last paper but the author would not be known best for it. Cork and Lawrence in modern alchemy make gold; unfortunately in "The Transmutation of Platinum by Deuterons." The process was not likely to influence the precious metals market. Later in the year, in Volume 50, we have an article (from a letter to A. H. Compton) by Arnold Sommerfeld on the shape of Compton lines; Beth describes fully his experiment on the detection of the angular momentum of light; Kellogg, Rabi, and Zacharias go into the "Gyromagnetic Properties of Hydrogen," capitalizing on an earlier paper in Volume 49 by Rabi "On the Process of Space Quantization"; their apparatus is now a double, inhomogeneous, magnetic, deflecting field arrangement. R. R. Wilson appears in his "Very Short Time Lag of Sparking" (order 10^{-8} sec), and Amaldi joins with Fermi in thirty pages from Rome "On the Absorption and Diffusion of Slow Neutrons." Gunnar Kellstrom from Sweden has his Letter on the viscosity of air that Birge referred to; Mitchell and Powers tell of their observation of the Bragg reflection of slow neutrons from a ring of crystals co-axial with the beam of neutrons, direct access of which to a detector is prevented by absorbing material (two cones of the material, base to base, the bases plane containing the ring of crystals around the outside); and Beams and Haynes at Virginia tell of isotope separation by centrifuging; Beams was big in high speed rotators. Cosmic rays are still a thriving activity; Anderson and Nedermyer report on their cloud chamber study, and Millikan and Neher have their "A Precision World Survey of Sea Level Cosmic Ray Intensities." But Compton seems to have settled matters as between photons and charged particles in the radiation with his "Cosmic Rays as Charged Particles" in Volume 50 (1936). The two men have had their confrontation, very much played up in the country press, at the famous A.A.A.S. meeting at Atlantic City, where the just returning Neher had to give the elder statesman the bad news that Nehr had found a latitude effect on his trip north from South America, having missed it in a set of circumstances on the way down. The dispute is described interestingly by Kevles in *The Physicists*. Millikan's reputation was not enhanced in the affair. An important investigation from Washington was also reported in Volume 50: Tuve, Hafstad and Heydenburg describe in a paper the scattering of protons by protons, and in a following paper, Breit, Condon and Present analyze the results and show them to correspond to S-wave scattering from a potential well representing a proton-proton attraction very nearly equal to the attraction between proton and neutron. In a Letter, Bothe and Maier-Leibnitz do a Compton scattering coincidence experiment and show the simultaneity in the scattering of the photon and that of the recoil electron. Bothe would in 1954 receive a Nobel prize for his work in the development of coincidence techniques.

While it may by now have been settled that cosmic rays were charged particles, puzzles remained. In Volume 51 (1937), Neddermeyer and Anderson consider in a "Note on the Nature of Cosmic Rays" whether there are possibly heavy electrons. A bit later at an APS meeting (Washington, April 1937, the 213th), Street and Stevenson submit an abstract also reporting on a penetrating component in the radiation that can't be electrons and can't be protons. Referring to these two reports, in the next REVIEW volume, E.C.G. Stueckelberg in Geneva, citing the prediction of Yukawa of "particles of this sort" and his own independent conclusion, believes the two groups have found a new elementary particle. Sadly, the Japanese also observed the heavy particle and sent their cloud chamber picture in to the REVIEW for publication, with text considered too long. So the editors sent it back and they missed out. In the volume's last entry, Oppenheimer and Serber comment on the particle in a Letter and point out Yukawa's suggestion and the possibility that this may be it. On the neutron front, Hoffman, Livingston, and Bethe offer "Some Direct Evidence of a Neutron Magnetic Moment" in scattering neutrons (polarized by passing them through magnetized iron) from a rotatable magnetized piece of iron (as analyzer), as suggested by Bloch in Volume 50. In 51, one Julian Schwinger at Columbia discusses the "Magnetic Scattering of Neutrons," and in a Letter,

Dunning, Powers, and Beyer, also at Columbia, do "Experiments on the Magnetic Properties of the Neutron," and Rabi, upstairs in the same laboratory, writes on "Space Quantization in a Gyrating Magnetic Field"; he will be doing some gyrating in the space between those two aforementioned inhomogeneous fields of his. Over at Pittsburgh, Stern is also at work with molecular beams and proposes measuring (Volume 51) the Bohr magneton through balancing by gravity the magnetic force acting on a beam, the analogue of Millikan's balanced oil drops. Mass spectrometry is making strides forward with Dempster's double focusing system, written up under the title "Electric and Magnetic Focussing Mass Spectrometer"; Bainbridge and Jordan had described a somewhat similar system, also involving energy and direction focussing, in Volume 50. Superconductivity is drawing attention: Slater writes on "The Nature of the Superconducting State" and Fritz London comments later with a Letter under the same title. In the Letter adjoining that, Zwicky considers "The Probability of Detecting Nebulae which act as Gravitational Lenses." Immediately preceding these two is one by Hideki Yukawa and Shoichi Sakata on K-capture. Yukawa would be remembered more for his meson, and would write at length in 1950 REVIEW papers, I and II, on "The Quantum Theory of Non-local Fields."

In the next volume of the REVIEW (52, 1937), we have a couple of notable solid state physics communications: Bardeen is into the "Conductivity of Monovalent Metals"—not yet superconductivity— and Conyers Herring has some aspects of energy bands in crystals— the effect of time reversal symmetry and accidental degeneracy. Estermann, Simpson, and Stern, at the Carnegie Institute of Technology, make some improvements in measurement of the magnetic moment of the proton; Elsasser from Zurich considers quantum measurements and the role of uncertainty relations in statistical mechanics; Wheeler writes on "Molecular Viewpoints in Nuclear Structure"; and Houston tightens up the viscosity of air, arriving at a result close to that of Kellstrom. In Letters, Bethe and Rose find a limit (because of relativistic increase of mass) to the maximum energy of a cyclotron beam, and Cerenkov from Russia sees "Visible Radiation Produced by Electrons Moving in a Medium at Velocities Exceeding that of Light." (Hark back to Franklin's insightful suggestion to Merritt forty years earlier, quoted up front in this chronicle.)

An important paper, "Note on the Radiation Field of an Electron," by Felix Bloch and Arnold Nordsieck shows how in field theory to avoid the infinity associated with the "infra-red catastrophe"—the emission of low frequency photons which tends to infinity in quantum theory; Nordsieck then uses the technique in calculating the radiative scattering of electrons in "Low Frequency Radiation of a Scattered Electron." Wigner, also in Volume 52, in "On the Consequences of the Symmetry of the Nuclear Hamiltonian on the Spectroscopy of Nucleii" introduces "isotopic spin" as label for Heisenberg's variable, which can have values +1 or –1 depending on whether it is assigned to the neutron or to the proton—considered as different states of the same particle.

With the first page in the lead issue of Volume 53 (1938), Ernest Merritt offers his last contribution to the PHYSICAL REVIEW. It is his memorial for colleague and REVIEW founder, Edward Leamington Nichols. A nice portrait accompanies the note. On the technical side, Zahn and Spees, following a Letter they had in Volume 52, measure "The Specific Charge of Disintegration Electrons from Radium E" in order to "distinguish between ordinary 'Lorentz' electrons and the widely differing special type of heavy electrons required by the speculation that the well-known beta-ray paradox might be explained by variations, with velocity, of the rest mass of electrons created in the nucleus rather than by the neutrino hypothesis." Later (in Volume 57), they do the experiment with positrons and find the same e/m_0 values, that which we prefer to see. In the 53rd volume, Clarence Zener has four papers on "Internal Friction in Solids," continuing what he had written in Part I in Volume 52; and Rose continues his discussion started in his previous Letter with Bethe on "Focusing and the Maximum Energy in the Cyclotron," half of which is something of a companion piece to the adjacent, well-known article by R. R. Wilson on "Magnetic and Electric Focusing in the Cyclotron." The Berkeley people were not very happy with any suggestion that there was a maximum limit in cyclotron acceleration, so that the Bethe–Rose contribution on the point was most unwelcome. They must have been relieved by two papers of L. H. Thomas, side by side, in the next volume (54). He shows that while the Bethe–Rose result may be true for a radially symmetric field, it is not true in general; for a field varying with polar angle, there is an additional focusing effect. He also treats the case of quadrants substituting for "Dees." Not that any machine incorporated such a complication until years later when Courant, Livingston, and Snyder, at Brookhaven, invented "strong focusing" for synchrotrons in a paper (Volume 88), followed by one of John Blewett, also of Brookhaven, on strong focusing in a linear accelerator. In fact, it was also invented in an unpublished manuscript by N. Christophilos, two or three years prior to the Brookhaven papers. The concept was first successfully implemented in a machine with Cornell's second synchrotron; although Christophilos' priority is acknowledged later (Volume 91) by Brookhaven, the laboratory is generally credited for the idea, seemingly somewhat unjustly. Actually it seems that Luis Alvarez at Berkeley was the first to try the idea, modifying offhandedly a linear accelerator; he showed the concept worked. High voltage breakdown limited the life of his modification to a few weeks. Altogether, a mixed bag.

Volume 53 had other significant work. Breit and Wigner discuss "The Saturation Requirement for Nuclear Forces" as do Critchfield and Teller; Sachs and Goeppert–Meyer make "Calculations on a New Neutron–Proton Interaction Potential;" Bethe contributes two pages on the "Binding Energy of the Deuteron"; and Gamow considers "Nuclear Energy Sources and Stellar Evolution." Teller and Wheeler write "On the Rotation of the Atomic Nucleus" and Crane and Halpern present pretty "New Experimental Evidence for the Existence of the Neutrino" in cloud chamber photographs of single beta decay events in radioactive chlorine, where the recoil is clearly seen with an unbalance in energy and momentum unless something else not recorded went also. Crane again, this time with Ruhlig, in a Letter has "Evidence for Particles of Intermediate Mass," in a sort of corollary to Neddermeyer's Letter on "The Penetrating Cosmic Ray Particles." In a landmark Letter, Rabi, with Zacharias, Millman, and Kusch, have installed Rabi's gyrating magnetic field in between the two focusing magnets to flip the moments of neutral particles passing through; great precision results. Their brief report is "A New Method of Measuring Nuclear Magnetic Moments" and a month later in a follow-up Letter, they have done $_3Li^6$, $_3Li^7$, and $_9F^{19}$. In a sense, the two magnets are like optical polarizer and analyzer, set parallel, with the gyrator between like a half wave plate at 45° so that at critical gyrator frequency the spin is flipped and the beam cannot make it through the second "analyzer" magnet. Really beautiful.

With Volume 54 (1938) we find two rather important papers by Luis Alvarez; one on the "Capture of Orbital Electrons by Nucleii," now known as K-capture, and the other on "Collimated Beams of Monochromatic Neutrons 300°–10°K"; he is modulating the plate voltage of the cyclotron oscillator to produce bursts of neutrons, the flight of which can be timed. A rather better method would come in 1941 (Volume 59) from Cornell where development and use was made of an arc source for the ions to be accelerated in their small cyclotron. Much greater ease of modulating the accelerated beam is to be had; a lot of slow neutron spectroscopy would follow in timing the flight of neutrons over a given distance with a bit of fancy electronics. C. P. Baker and R. F. Bacher describe the work in their 1941 paper "Experiments with a Slow Neutron Velocity Spectrometer." Back in 1938, however, Alvarez was blazing trail. On the other hand, Cornell did make a major contribution in the year with Livingston and M. G. Holloway's "Range and Specific Ionization of Alpha Particles" and with Williams' two papers: one on e/m from the H_α–D_α interval, and the more important one with his final word on the subject of the hydrogen (deuterium) fine structure; theory definitely seems to be wrong. The Letter by Pasternack suggesting that the $2S^{1/2}$ level should be moved about 0.03 cm^{-1} comes later in the volume. Linus Pauling writes from Cal Tech on "The Nature of Interatomic Forces in Metals"—work stimulated by Arnold Sommerfeld—and Otto LaPorte writes on that hypothetical particle dreamed up in Japan and Geneva, presumed to provide the force between proton and neutron: "Scattering of Yukawa Particles"— "Dedicated to Arnold Sommerfeld upon His Seventieth Birthday." In triggered cloud chamber pictures, Neddermeyer and Anderson find "Cosmic Ray Particles of Intermediate Mass," of about 240 m_e. And in this year, 1938, Heitler submits some "Remarks on Nuclear Disintegrations by Cosmic Rays," from H. H. Wills Laboratory in Bristol. Bethe and C. L. Critchfield have an important paper on "The Formation of Deuterium by Proton Combination," a process discussed a year earlier by Weissacker; they calculate about the right order of magnitude for the evolution of energy in the sun. A significant article by London, "On the Bose–Einstein Condensation" and another one by E. M. Purcell extending the Hughes–Rojansky cylindrical electrostatic focusing to "The Focusing of Charged

Particles in a Spherical Condenser," suggested years earlier by Aston, are both harbingers of more memorable things to come.

With the year 1939 we see the beginnings of physicists' loss of innocence. The last half of the decade had seen the Nazi rise to nightmarish ascendency and the flight of many of Germany's most gifted people to countries more hospitable, to put it mildly. Scientists left in large numbers; the United States was among the beneficiaries and we have seen some of their names in the foregoing. The world was bracing for war; it would come in this year and the contributions from physics would change the world forever. In the PHYSICAL REVIEW, however, things are about as usual. In Volume 55 of the year, there is the landmark, Nobel recognized (1967), paper by Hans Bethe, "Energy Production in Stars," which follows his earlier Letter of the same title and his paper with Critchfield in the previous volume. The carbon-nitrogen cycle is seen as appropriate for heavy, bright stars on the main sequence while the proton-proton cycle is more suited to light, fainter stars with lower central temperature; the sun fits either tolerably well. Alfven considers the motion of cosmic rays in interstellar space; Willard Libby and D. D. Lee (not T. D.) write on the energies of soft beta radiation and method of their determination; Harrington is concerned unhappily that his viscosity of air measurements, on which Millikan has relied, seem to be wrong some way—where error could have arisen he cannot see; but that constant is still a bother. Crane attempts in a rough test to observe the absorption of neutrinos in the reaction $Cl^{35} + \mu \rightarrow S^{35} + e^+$ followed by $S^{35} \rightarrow Cl^{35} + e^- + \mu$, anticipating experiments going on today. He didn't see anything. His μ has become symbol ν to particle physicists. There is a Letter by Vallarta and R. P. Feynman— "The Scattering of Cosmic Rays by the Stars of a Galaxy," and one from Japan on Yukawa's particle by Nishina, Takeuchi, and Ichimiya (a nice set of names), "On the Mass of the Mesotron," analyzing a cloud chamber track and coming to a mass of 180 m_e. In this year, Alvarez and Cornog at Berkeley do the "first significant experiment using the technique of accelerator mass spectrometry" in identifying He^3, reported in a REVIEW Letter of Volume 56; radioactive H^3 followed the discovery in their bombardment of deuterium with deuterons (Volume 57), finally clearing up the uncertainty in that isotope.

But these Letters are as nothing compared to the flood which came in after mid-February of 1939. Bohr had arrived in the country bringing sensational news from Europe, sending our nuclear physicists in haste to their laboratories and calculators (gear driven). From the Department of Terrestrial Magnetism, Roberts, Meyer, and Hafstad lead the parade with "Droplet Fission of Uranium and Thorium Nucleii." Green and Alvarez report from Berkeley on "Heavily Ionizing Particles from Uranium"; Fowler and Dodson at Hopkins on "Intensely Ionizing Particles Produced by Neutron Bombardment of Uranium and Thorium." There is another Berkeley contribution, an article by Abelson—"Cleavage of the Uranium Nucleus," finding K_α X-rays from iodine; and one by Bohr himself, from Princeton, on "Resonance in Uranium and Thorium Disintegrations and the Phenomenon of Nuclear Fission," as much as anything, written to protect the priorities of European colleagues according to Rhodes in his *Making of the Atomic Bomb*; the information had inadvertently gotten away from him before it was intended. There were more Letters in the next issue: Corson and Thornton; McMillan; Roberts, Meyer, and Wang; Anderson, Booth, Dunning, Fermi, Glasoe, and Slack. (Fermi had by now been safely in this country only a short while, at Columbia, having made the one-way side trip to the U.S. in his junket taken to Sweden for reception of his Nobel prize. The Letter details briefly a bit of the discovery.) Then comes an ominous and prescient Letter from Anderson, Fermi, and Hanstein, "Production of Neutrons in Uranium Bombarded by Neutrons"; another by Szillard and Zinn on "Instantaneous Emission of Fast Neutrons in the Interaction of Slow Neutrons with Uranium"; and yet another by Dunning, Booth, and Slack, "Delayed Neutron Emission from Uranium." And still more Letters: Abelson with "The Identification of Some of the Products of Uranium Cleavage"; Kennedy and Seaborg in a search for beta particles in the breakup; Anderson and Fermi on "The Simple Capture of Neutrons by Uranium"; and Libby on "Stability of Uranium and Thorium for Natural Fission." The noteworthy report of Rabi's group on their beautiful work on the moments of $^3Li^6$, $^3Li^7$, and $^9F^{19}$ was almost lost in the excitement.

By summer, things had cooled a bit, at least in the REVIEW. In Volume 56, Wigner, Critchfield, and Teller consider "The Electron–Positron Theory of Nuclear Forces," which Teller and Gamow proposed earlier (Volume 51, 1937) as a way of explaining the equality of proton-proton and proton-neutron forces; V. F. Weisskopf, by now in this country, writes "On the Self Energy and the Electromagnetic Field of the Electron," proving that the self-energy is (only!) logarithmically infinite in the positron theory "to every approximation in an expansion of the self-energy in powers of e^2/hc." Oppenheimer and Snyder have their well-known paper, "On Continuing Gravitational Contraction"; the black hole is born. Townes and Smythe, at Cal Tech, determine spectroscopically "The Spin of Carbon Thirteen"; Lester Germer has a nice piece with beautiful transmission electron diffraction pictures in "Electron Diffraction Studies of Very Thin Films"; and R. P. Feynman considers "Forces in Molecules." But fission is much on people's minds. Bohr and Wheeler get their liquid drop model in "The Mechanism of Nuclear Fission," and Anderson, Fermi, and Szillard in a short paper write of "Neutron Production and Absorption in Uranium." They are looking ahead when they write, "... it is of interest to ascertain whether and to what extent the number of neutrons emitted exceeds the number absorbed." Further: "From this result we may conclude that a nuclear chain reaction could be maintained in a system in which neutrons are slowed down without much absorption until they reach thermal energies and are then mostly absorbed by uranium rather than by another element." A diagram of their experimental arrangement already looks like a small pile. Zinn and Szillard in "Emission of Neutrons by Uranium" decide that about 2.3 neutrons are produced per fission on average.

By Fall, the subject was much less reported on; we had such a paper as "Emissive and Thermionic Characteristic of Uranium" by Hobe and Wright, having nothing to do with fission. Bohr and Wheeler had a Letter on "The Fission of Protactinium by Fast Neutrons," agreeing with the observation at Columbia that slow neutrons did not cause fission in the element, but who emphasized the importance of doing experiments on separated isotopes of uranium. It is interesting to follow the indices of the REVIEW for the years; all that early flurry of work and then relative subsidence for a while. There comes Bearden's long paper on his precision measurement of the air viscosity, a very careful and impressive piece of work, which nailed down finally that troublesome constant. The oil drop "e" was at last brought into accord with that determined via X-rays. On another front, Burton, Hillier, and Prebus give a report on "The Development of the Electron Microscope at Toronto," and R. W. Wood has his swan song with that lecture demonstration of resonance radiation, swan song save for two meeting abstracts, one on large gratings for Schmidt telescopes and one on artificial meteors.

In Volume 57 (1940, Europe was facing invasion of the lowlands in the "phony war"), the death of another elder physics-statesman was noted with a portrait and memorial statement on Floyd K. Richtmyer of Cornell. And from Cornell, Bethe has an article in two parts on the "Meson Theory of Nuclear Forces"—"I: General" and "II: Theory of the Deuteron." Pomerantz at the Franklin Institute in an article, "The Instability of the Meson" and Rossi, Hilberry, and Hoag on the "mesotron disintegration" arrive at a lifetime for the particle of about 2×10^{-6} sec, improved by Rossi's group in 1942 to about 2.2×10^{-6} sec, today's accepted value. Holloway and Bethe check a last unknown value needed in Bethe's stellar reactions, in determining the cross section for $N^{15}(p,\alpha)C^{12}$, and Conyers Herring comes up importantly with "A New Method of Calculating Wave Functions in Crystals." Victor Hess, cosmic ray Nobel Laureate, has an article on seasonal and atmospheric temperature effects on "his" radiation. Podolsky and Branson in "On the Quantization of Mass" use Dirac's equation and a space-time manifold of Eddington's; taking the electron mass we know, leads to a universe radius of 10^{10} cm, or taking the universe radius as 10^{28} cm (believed about right), leads to an electron mass of 10^{-65} gm. Something wrong. Bloch and Siegart discuss "Magnetic Resonance for Non-rotating Fields," which would be useful to Bloch and others in time. In fact, it was useful already to Alvarez and Bloch in the same volume in their paper on the magnetic moment of the neutron, in which they treat the neutron to a variant of Rabi's polarizer-spin flip-analyzer arrangement; instead of inhomogeneous fields as polarizer and analyzer, they make use of the scattering from magnetized iron as polarizer and analyzer, as did Hoffman *et al.*, before. Very pretty; there sure are some clever people in this business. The Rabi apparatus itself was used in an important measurement at Columbia by Kellogg, Rabi, Ramsey, and Zacharias; they determine the electrical quadrupole moment of the deuteron. To wind up the volume, Bethe has a long article on "A Continuum Theory of the Compound Nucleus," not to mention a shorter one with Nordheim, "On the Theory of Meson

Decay." And Fermi worries about ionization loss of energy of fast charged particles in gases and condensed matter.

In the REVIEW. index for Volume 57, there is a section given to Fission of the Nucleus—a total of only eight letters, two articles, and three meeting abstracts in the whole volume. Among the entries: an article by Kanner and Barschall on the "Distribution in Energy of the Fragments from Uranium Fission," and one by Dodson and Fowler, "Products of the Uranium Fission—Radioactive Isotopes of Iodine and Xenon"; a Letter by Louis Turner, who arrives at 2.6 for the number of "Secondary Neutrons from Uranium," agreeing with French workers that no explosive reaction can occur in any mixture of uranium and hydrogen; an important result is contained in a Letter from Dunning, Booth, Grosse, and Nier on "Nuclear Fission of Separated Uranium Isotopes." In his mass spectrograph at Minnesota, Nier has separated a minute amount of U^{235} from natural uranium; the work shows that the isotope is responsible for the slow neutron fission, as Bohr and Wheeler had predicted in their model. A second Letter from Nier and his Columbia collaborators, and one from G. E. by Kingdon and Pollack only strengthen the observation. An unrelated Letter, from Copenhagen, is by G. Hevesy defending against some criticism about his technique of using radioactive tracers in the study of physiological studies. He comes off all right; in 1943 he would receive the Nobel prize in chemistry for just such work.

In Volume 58 (1940) we have a Letter by Donald Kerst on "The Acceleration of Electrons by Magnetic Induction," a trick others had tried but failed of doing. Gamow and Schoenberg have a Letter on "The Possible Role of Neutrinos in Stellar Evolution," particularly in stellar collapse as a supernova (elaborated on in their Volume 59 article and stunningly observed in 1987 with the bright Magellanic cloud object, SN-1987a). Marton, now at the RCA Laboratories, has "A New Electron Microscope," achieving 50 Å resolution on biological objects. Zurich's W. Pauli, writing as a European "escapee" in Princeton, has an important application of special relativity in "The Connection Between Spin and Statistics," and Schwinger and Corben also theorize on "The Electromagnetic Properties of Mesotrons," following a one-page comment from Moscow by Ig. Tamm (1958 Nobelist with Cerenkov) on "Mesons in a Coulomb Field," and work of Oppenheimer, Snyder, and Serber in Volume 57. From Japan,

Tomonaga and Araki write on the "Effect of the Nuclear Coulomb Field on the Capture of Slow Mesons"—curious particles then, and today still curious. Fission continues with us, the Index section of our journal somewhat larger than in the first appearance, a bit of resurgence following the lull after the initial burst of activity. From Copenhagen, Brostrom, Boggild, and T. Lauritson have "Cloud Chamber Studies of Fission Fragment Tracks," on which Bohr comments in a separate article and, joining with the team, in "Velocity-Range Relation for Fission Fragments." Additionally, he considers "Successive Transformations in Nuclear Fission" in another paper. Willis Lamb has a theoretical paper on the "Passage of Uranium Fission Fragments through Matter." The Columbia group and Nier have another of their Letters on separated isotopes with "Neutron Capture by Uranium 238," also important like their work with U^{235}. One of the shortest Letters in REVIEW. history, if not *the* shortest, but in its way perhaps as significant as most, was the nine lines in Volume 58 from Flerov and Petrjak in Russia on the rare "Spontaneous Fission of Uranium." Herbert York, one-time big wheel in American science policy and a Director of Livermore Laboratory, indicates (*Physics Today*, April, 1988) that the lack of any American response to the note was the early tip-off to the Russians that we were embarked on a program to put fission to practical (impractical?) purpose. There were obviously other clues to be had.

In Volume 59 (1941) Baker and Bacher tell of their slow neutron spectroscopy, and Gamow and Schoenberg finish their stellar collapse with an article on the neutrino role. Marshak and Weisskopf are together at Rochester with "On the Scattering of Mesons of Spin $h/2$ by Atomic Nucleii."

In Volume 60 (1941) of our journal, Kerst describes fully his work with the induction accelerator, dubbed the "Betatron," Goeppert–Meyer is present in her "Rare Earth and Trans-uranic Elements," and Kapitza ("captured" and held in the USSR) manages to get through to us with "Heat Transfer and Superfluidity of Helium II" (he would be awarded the Nobel prize in 1978 for his work on the helium liquifier), is followed by Landau, also managing to get through, with the "Theory of Superfluidity of He II." That substance will be a puzzle for quite a while. Millman and Kusch determine nuclear moments in a direct comparison with the electron moment, and

Melba Phillips, in an accompanying companion piece, worries for them about possible deviations from $g_j = 2$ for the electronic ground state of the alkalies, assumed to be exactly two by Millman and Kusch; thereon hangs another tale. Raman and Nilakantan have a Letter on the "Classical and Quantum Reflections of X-rays," and Birge is still at his forte with new values of the Rydberg constant and e/m obtained from the hydrogen, deuterium, and helium spectra. One of the seminal works in lattice physics comes in two papers of Volume 60 (Parts I and II) by H. A. Kramers and G. K. Wannier: "Statistics of the Two Dimensional Ferromagnet." The work has led to much progress in related fields; in particular, Lars Onsager's exact solution to the two-dimensional Ising model, with which so many discussions of magnetism and phase transformations seem to begin, was an outcome. That important, long paper would come in a few years, during the war, in Volume 65 (1944) as "Crystal Statistics. I: A Two Dimensional Model with an Order-Disorder Transition." II seems never to have been written and clearly was not necessary. Fission is present, but at low level; Bohr has a Letter on deuteron-induced fission, and Fermi and Segre one on fission induced by alpha particles, not a process of high priority; there is a significant paper from Rome by other Italians, who are holding the fort, Ageno, Amaldi, Bocciarelli, Cacciapioti, and Trabacchi (what a fine string of names!) on the "Fission Yield by Fast Neutrons," and Boggild writes more on the cloud chamber work there in Copenhagen. And that is about it; fission goes underground.

Not that there were no more papers written. Turner at Princeton, stimulated by a *New York Times* article on the slow neutron fission of separated U^{235}, writes a paper, "Atomic Energy From U^{238}," suggesting that successive trans-uranic elements made in the irradiation of U^{238} by neutrons might themselves fission. Submitted to the REVIEW in May of 1940, it never saw the light of day until after the war in Volume 69 (1946), accompanied by a number of Letters from Berkeley pertinent to one of those trans-uranic elements, plutonium, element 94: Seaborg, McMillan, Kennedy, and Wahl on "Radio-active Element 94 from Deuterons on Uranium" (Jan. 1941) with a follow-up Letter of the same title by Seaborg, Wahl, and Kennedy in March and a "Search for Spontaneous Fission in 94^{239}" by Kennedy and Wahl in December.

The matter of secrecy in the field of fission reared its ugly head early on and the Physical Review would soon see the effects. Within a short period after receiving Bohr's news of fission, both Fermi and Szillard had written reports on the secondary neutrons from fission ready to go to the Review. Pegram, elder statesman, also at Columbia, counseled going ahead to establish their priorities, suggesting submitting the Letters but that publication be held up until the issue of secrecy was resolved. Teller and Szillard both were of the opinion that there should not be publication; Fermi was of the opposing view. The particular subject became moot with *Nature's* publication in March of 1939 of the Letter of von Halban, Joliot, and Kowarski on neutrons from fissioning uranium. (See Rhodes, *The Making of the Atomic Bomb.* Some of the fission business here is culled from his fascinating book.) There were differing opinions on the secrecy matter. The seeming relative subsidence of activity after the first flurry was indeed somewhat the result of voluntary self-censorship. The story is told by Spencer Waert in an also fascinating account in *Physics Today* (February 1976): "Scientists with a Secret".

In 1942 we in America are in the war; Fission of the Nucleus has disappeared from the Index, volumes of the Review get sharply thinner; it comes bi-monthly, two months' worth at a time, but little enough. John T. Tate is listed as Editor (On Leave); his colleague Buchta acts in his stead. Nonetheless, there is some publishable and good physics to report. In a Letter of Volume 61, Kerst has his machine in a new version at G. E. up to 20 MeV, Rossi and his crew polish up their mesotron life-time measurement, and Millikan (forever) is still into cosmic rays, worrying in this volume and the next with Neher and Pickering about their origin; at least it is now agreed that they are charged particles. From Scotland, and away from Nazi Germany, Max Born makes a brief appearance with a comment on temperature diffuse scattering of X-rays. In Volume 63 (1943) Shull, Chase, and Myers, observe polarization of electrons in the scattering from thin foils, and Jauch writes on the "Meson Theory of the Magnetic Moment of the Proton and Neutron." Yardley Beers has a rather nice paper on "Direct Determination of the Charge of Beta Particles"; they are electrons all right. In 1944, Volume 65, Onsager highlights the year with his notable contribution noted above; Lande and Thomas have a third paper on "Finite Self Energies in Radiation Theory: III," I, preceding, coming from Lande in Volume 60, and II, with Thomas, in the same volume. Cobas and Jauch have a paper on the "Extraction of Electrons from a Metal Surface by Ions and Metastable Atoms"; important to the understanding of gas discharge currents. The last item in the volume is a letter from Russia; significant for things ahead, Iwanenko and Pomeranchuk consider radiation from high energy electrons in circular orbit in "On the Maximal Energy Attainable in a Betatron." Such radiation would in time become very useful and a major by-product of high energy machines. In Volume 66 (1944) Millikan and company have more on cosmic ray origins (understandably, Millikan was not the power that he was in the First World War, so that in the Second War he could worry about such matters), and Bethe has a lengthy paper on "Theory of Diffraction by Small Holes," a contribution he could make for the Radiation Laboratory at MIT before he was cleared for more important work taking him to the Southwest. In Volume 67 (1945) there are a few conspicuous papers: John Blatt on the "Meson Charge Cloud around the Proton"; Jauch on "n–p Scattering and the Meson Theory of Nuclear Forces"; and Brillouin on "A Theorem of Larmor and Its Importance for Electrons in Magnetic Fields," applying it to electron beam flow in longitudinal magnetic fields, made somewhat restrictive by J. R. Pierce in a Letter of Volume 69. Kron at G.E. develops "Electric Circuit Models of Schrodinger's Equation" and uses the hardware with Carter in an "AC Network Analyzer Study of Schrodinger's Equation." Volume 68 (1945) of our journal is the thinnest PHYSICAL REVIEW in its near one hundred year history; it is all of 290 pages, but for all of that, includes some noteworthy items: McMillan's "Synchrotron—a Proposed High Energy Accelerator" (also dreamed up independently by Veksler in Russia) and his "Radiation from a Group of Electrons Moving in a Circular Orbit." Both aspects—the machine and its radiation—have become big physics. In the volume is another Letter, this from Ruark, on the atom of positron and electron—positronium. Hamilton, Heitler, and Peng, isolated in Dublin, have a long paper on cosmic ray mesons, and Cook finds K^{39}/K^{40} abundance ratio the same for rocks of different geological age in mass spectra analysis. Finally, a paper on the "Meson Intensity in the Substratosphere" by India's well-known high energy physicist, H. J. Bhabha, with Aiya, Hoteko, and Saxena, and a Letter on the brilliant (*Brighter than a*

Thousand Suns) result of all that fission business: Coven reports his accidental observation of the Almogordo blast.

In 1946, the war over, with Volume 69, the REVIEW started its recovery, getting fatter, rapidly fatter, in the flood of war work accumulated and declassified, and in the ideas conceived and stashed away during the war years. The journal's new growth was another sort of explosion, which has hardly abated to this day; 200 feet of shelf space! We are surely not going to try perusing that volume by volume.

Chapter 11
A BRIEF LOOK AT THE SECOND HALF CENTURY AND BEYOND

•◆•

In 1950, the PHYSICAL REVIEW in Volume 79, gave "its pages to John Torrence Tate, who for nigh on to a quarter of a century gave them to the physicists of the United States and the world." While suffering a stroke a few months before his death, Tate was still editing the two APS journals, the REVIEW and the *Reviews of Modern Physics*. However, following the loss, Minnesota would still have management of the journals for another year on an interim basis under E. L. Hill and John Buchta, after which S. Goudsmit and S. Pasternack would take on the responsibility of editorship at Brookhaven, with a doubled Editorial Board. From 1931, Buchta had served with Tate as an assistant editor, Hill stepping in also as occasion warranted. Buchta, like Tate, was a product of Nebraska, earned a B.S. degree in Electrical Engineering and a Master's degree in physics before migrating to Minnesota for his Ph.D. He remained there for the rest of his career, becoming Chairman of the Department (for long a strong group). He received the A.A.P.T. Oersted medal in 1937. During World War II, when Tate was gone and involved with the government in the war effort, Buchta served as the REVIEW Acting Editor. It was a trying time for a journal of physics when everyone was into war-related work, with little time to devote to non-classified physics and to the writing of articles for the REVIEW. But we came through. Following the war, and with the tremendous growth in physics and in its journals, the vicissitudes of changes and managements of the REVIEW got so complicated that further tracking of the editorship will be pursued no further herein.

It is difficult enough to keep track of the volume numbering and varieties (!) of the REVIEW, let alone its editors. In 1929, as we have noted earlier, the journal started coming every two weeks with Volume 34 of Series II, and (except for the anomalous year 1932) continued at that rate until the time of World War II. During that lean, war period some issues were for two whole months, and even so, the volumes got very thin; that for 1945 had only about 300 pages, large enough to be surprising at that, considering the state of the world. And then, with the war over, things blew up, successive volumes rapidly becoming very much thicker. By 1960 some single issues were running to 700 and 800 pages—every two weeks 700 more pages!

Probably the most significant postwar development to the journal was the establishment of the PHYSICAL REVIEW *Letters*, in July of 1958, sixty-five years after the REVIEW itself first appeared. The *Letters* section of the REVIEW had become so well used and publication times slow enough that a separate enterprise dedicated solely to fast letters was deemed appropriate. Goudsmit removed *Letters* from the REVIEW and instituted the new supplement. George L. Trigg, on a year's leave from Oregon, started it off and returned three years later to remain as Editor for thirty years (see PRL, July 1, 1988, "Farewell,—"). The supplemental journal was started, quite as an experiment, as editors Goudsmit and Trigg editorialized in the first of the bi-weekly issues. Speed of publication "at the expense of printing elegance" was a prime aim. Volume One of *Letters* held nearly 500 pages, supporting the aim. As with the first Letter in the PHYSICAL REVIEW back in 1930, there is perhaps some question as to how necessary for speed was the first communication to PHYSICAL REVIEW *Letters*, "Magnetization in Single Crystals of Some Rare Earth Ortho-ferrites" by Bozarth *et al.* at Bell Laboratories. The last one of this first issue, 40 pages later, was perhaps more like it—on the neutrino rest mass, by Sakurai. In between were matters of condensed matter physics, nuclear physics and particle physics; there was a note from Los Alamos on neutron scattering by liquid helium, resulting in a pretty $E(k)$ *versus* k plot with the nice roton minimum; and one by Telegdi *et al.* on the precise measurement of the muon ($\pm$) magnetic moments by a "stroboscopic coincidence" method. Publication time was fairly fast—about three weeks. Useful also in the *Letters* were the abstracts printed with each issue, of papers to be published in the full blown REVIEW itself.

Ten years later the editors were not sure the experiment was a success. In 1964 the journal had become a weekly; so, yes, technically it was a success. That could be inferred as well by the number of other journals which, following the example, had also started Letters supplements. The need was there. The REVIEW intent had been to have rapidity in publication "of just those reports that reasonably might be expected to have immediate impact on the research of others." That they also be of general interest came along later. The *Letters* reports during the first decade had come rather to have a prestigious value which the editors felt was not deserved; they remarked in agreement on another journal's editor commenting that "urgency was not necessarily equivalent to importance." Nonetheless, the "experiment" was continued and today no physicist would have it discontinued. For the first decade, no abstracts accompanied the communications; at least they were few and far between. In 1969 the instructions to contributors indicated the previous "complete" optional abstract was now to be required, and so it has been since.

In the first ten years, REVIEW *Letters* expanded, somewhat like its parent journal, at the same time maintaining rather well its character of general interest, which one hopes can be continued. The editors pretty much managed to curtail unbridled growth of the supplement. Unbridled is said somewhat reservedly; in the tenth year, 1968, Volume 21 held over 1500 pages, already up a factor of four over the first volume. In that same time span, and for a like six-month period, the PHYSICAL REVIEW went from 2200 pages in 1958 (Volumes 111 and 112) to nearly 14,000 in 1968. By 1993 the number of submissions was growing by about 9.5% a year. The size of REVIEW *Letters* for that year had gone to 8800 pages, while the PHYSICAL REVIEW had gone during the same period to a total of nearly 70,000 pages! Each issue of REVIEW *Letters* is still reasonable in size and content; one can still look forward to its arrival, much as was the case with the pre-war REVIEW. This has meant that the editors are having to exercise more and more severe judgments on the suitability of the submissions; fewer and fewer qualify as of sufficient "general" interest. *Brief Reports and Rapid Communications* to the appropriate A, B, C, D or E PHYSICAL REVIEW would take care of those very special interests. The time between manuscript arrival and time of publication in the *Letters* is closer to a few months rather than the early three weeks, except in very unusual circumstances (for example, sensational discovery); but it all works surprisingly well.

In a survey article in the November 1981 issue of *Physics Today* (a large and memorable number) on the occasion of the 50th anniversary of the founding of the American Institute of Physics, Norman

Ramsey also contrasted the size of the PHYSICAL REVIEW and its supplemental *Letters*, in 1981 as opposed to the situation fifty years earlier. The 1931 date is also an appropriate date in the REVIEW's progress. Spencer Weart in the same A.I.P. anniversary issue, noted that "In 1931 the PHYSICAL REVIEW for the first time was cited more often in physics literature than its chief rival, the German *Zeitschrift für Physik*." Ramsey found that the December 1931 PHYSICAL REVIEW contained 22 papers while that of December 1981 had 350. Letters in the REVIEW of that early date numbered eleven; in December 1980, PHYSICAL REVIEW *Letters* had 106. This had meant, of course, what with the increase in number of papers submitted to REVIEW *Letters* that there has been an increase in the delay time of publication. Ramsey notes that it was about 21 days in 1931 and had gone to 138 days in 1980, and he extrapolated that it will be about 700 days in 2031, solving automatically the publications problem!

In a "guest" commentary in *Physics Today* (April 1988) Cornell's Professor Mermin raised question as to whether anyone was reading the *Letters*; it is well agreed that the PHYSICAL REVIEW was beyond perusing, let alone reading. It had come to be pretty much archival in nature. Mermin was concerned that *Letters* had come to about the point of usefulness and size that he had found the REVIEW to have when he was a graduate student, somewhat before it had become known to him and some of his colleagues as the "green plague." (It was certainly never known as *that* at Cornell!) And he wondered if it was being read at all. He had found a misspelling—Lagrangean—occurring in the *Letters* extending back at least a few years. No one had spotted the error in that period, and certainly physicists knew how to spell Lagrangian; he had sampled that in a private poll. He raised the important point, not previously mentioned in this memoir, that there is a large, informal circulation of preprints in operation; physicists see to it that their results get distributed to a large number of those engaged in similar work, obviating the need to read other journals. Mermin was finding his office becoming engulfed in preprint literature! (So perhaps even the preprint trade is out of hand.)

Another outside comment came in mid-1989 with editor John Maddox writing in *Nature* (July 13), bemoaning an aspect or two touched on in this review of the REVIEW. He must have had a bad morning. He notes the instructions given to authors in the backs of journals, particularly in PHYSICAL REVIEW *Letters*, "among the most authoritarian of all journals"—"dogmatic on space contributors may fill"—standard recipe by which they "can calculate for themselves how many keystrokes of their typists fingers will fill up a maximum of four pages of text." But the personal concern of Goudsmit as editor made it "among the most authorit*ative* of journals." Of his contributors he had one complaint: they liked too well to publish "on the front pages of the *New York Times* as well as in PRL." In the 1960's, "there was always an anti-proton or some other exotic particle whose discoveries seemed that important." Goudsmit's war on the "practice of prior release" kept contributors "well versed in explanation of how their work had arrived there by accident." But now, Maddox notes, the *Letters* policy "includes a statement on prior disclosure that must have even unshockable Sam Goudsmit turning in his grave." APS journals require that "authors should not previously have published what they have to say elsewhere." Maddox comments on the silence, however, regarding "the pre-print business," of Mermin's concern. (We have seen herein how duplicate publication did not greatly bother the first REVIEW editors.) The PRL instructions repeat the familiar exceptions to the rule on prior publication wherein the discovery is reported "first in popular magazines"—for example, *Time*, *Newsweek*, *Scientific American*, and *Physics Today*,— "but prior publication by 'newspaper, television, and radio' is not held to be a restraint of publication" in an APS journal! Thus, Maddox complains, proper public recognition of a discovery comes to depend too much on the public relations enterprise of institutions—"and even those in universities are not above reproach." (Utah comes to mind these days.) "Sam Goudsmit, being the first, may have been an unsystematic editor of PRL, but his instincts on this issue were correct," Maddox concludes.

The editorial staff of the PHYSICAL REVIEW considered the *Nature* editorial as quite inaccurate and outrageous; Maddox just did not have his facts straight. Goudsmit was very much around when the present position was taken in 1976; it was seen then to be unseemly for a researcher who has been supported by the public, to hold results back from the public which the public had paid for, during the months taken to get the news out in PHYSICAL REVIEW *Letters*. Goudsmit's earlier prohibition, admittedly arbitrary, and at that somewhat selective, is current policy of the *New England Journal of Medicine* but apparently not that of *Nature*, to judge from its reports of the Utah hubbub. Notwithstanding possible institutional manipulation in early announcement to the media of hot discovery, and of some opinion in the Physical Society agreeing with Goudsmit's early position, it is still viewed as unreasonable by the APS Council, and Publications Committee, which recently reasserted the current policy deplored by Maddox.

In 1970 the possibility appeared of some relief from the flood in the PHYSICAL REVIEW. From the journal offices came another supplement: PHYSICAL REVIEW *Abstracts*, in which for "the community and related researchers" was (and is) published abstracts of coming papers in the PHYSICAL REVIEW (publication of which abstracts up to then had been a feature of the REVIEW *Letters*), the PHYSICAL REVIEW *Letters*, and the *Reviews of Modern Physics*. *Science Abstracts* had long since become unmanageable for casual going through; for a search of papers on a specific subject, by a specific author, or, indeed, hunting for a long lost paper, that compendium of abstracts was (and is) invaluable. Of course, it was no longer part of the journal package sent to Physical Society members; that went by the boards long ago. In fact, no longer does the American Physical Society even involve itself in the support of that great publications endeavor. So the PHYSICAL REVIEW *Abstracts* does serve the useful purpose of allowing readers to scan and find out what is appearing in the three APS journals—which are copious enough—more easily than plowing through them individually. It thus serves in a very restricted way (how could it be otherwise these days?) the purpose of the early *Science Abstracts*.

(Up to this point in our survey, when individual papers have been cited, it has not been necessary to identify the journal or source as either the PHYSICAL REVIEW or the PHYSICAL REVIEW *Letters* when referring to such and such a volume of this or that year; only the PHYSICAL REVIEW has been involved.) While the REVIEW *Letters* did not come on until 1958, we will from here on out (1945—) specify the PHYSICAL REVIEW or its *Letters* in the citation of specific papers; in the interest of a little variety and readability, the full journal title will not always be spelled out.

While PHYSICAL REVIEW *Letters* relieved the parent journal of a little of its reporting, it did not noticeably stem its growth (to put it mildly), which was not the intent anyway. In 1964, with Volume 133, the REVIEW was split into two sections: Section A—"Solids" and Section B—"Nuclear," each arriving biweekly on an alternating basis: Section A every two weeks in between issues of Section B; a conscientious subscriber could every month get four full issues, a new volume number every three months. This "happy" solution lasted but two years. With Volume 141, in 1966, there was a new *volume* every month (!), the volume number being followed by a superscript: 1, 2, 3, or 4, every week successively. Volumes 1 and 2 were devoted largely to solids and Volume 3 and 4 to nuclear and particle physics. A year later the superscripts had been increased to 5, the first three devoted for the most part to solid state or other condensed matter physics; every five days after the first of each month a new issue out; the lucky printers had a five or six days respite at month's end. And that continued for but three years when the new Series and present system was instituted. Confusion aplenty and consternation in the publications front office, not to mention for the subscribers. So now in this new Series III, we have five different PHYSICAL REVIEW's: A—Atomic, Molecular, and Optical Physics, coming monthly; B— Solid State Physics (changed to Condensed Matter with Volume 18 in 1978) coming biweekly and in two Sections, I and II (in the month's second mailing); C—Nuclear Physics, coming monthly (nuclear physics is fading); D—Particles and Fields, coming biweekly, and the newest PHYSICAL REVIEW, PHYSICAL REVIEW E, a spin off from PRA, E—Statistical Physics, Plasma, Fluids, and Related Interdisciplinary Topics, coming monthly. Each issue comes with its own *Brief Reports*, *Rapid Communications*, Comments, and *Addenda*, and of course, *Errata*. In a way, the Rapid Communications is a sort of PHYSICAL REVIEW *Letters*, but rather more specialized and not quite so limited as to length. Finally, there are indices. After the 1920 Cumulative Index, including all of the REVIEW Series I, there was not another Cumulative Index until that of 1950, going back to 1921, Authors and Subjects, a rather thick volume. Thereafter, an Index appeared every five years, listing authors only and, starting with that of 1955–1960, including authors of reports to PHYSICAL REVIEW *Letters*. The growth in the size of the Indices is spectacular. Beginning in 1970, we have yearly indices, thick enough, covering both the PHYSICAL REVIEW and PHYSICAL REVIEW *Letters*, by subject and by author, with cumulative author indices coming periodically (six, three, five years—the six year number (1960–1976) impressive indeed). *In toto*, an enormous operation.

Further to emphasize the magnitude of the operation, we may note that the American Physical Society Editorial Office out near Brookhaven (Ridge, N.Y.) has over 100 full-time people in just editorial work and PHYSICAL REVIEW *Letters* production activities, with another ten or so APS people supervising some 100 employees of the Institute at its Woodbury, N.Y., Center, where a total of over 350 people labor. There are a few dozen part-time employees scattered about elsewhere and some ten thousand (!) referees used each year by the Society alone. In 1989, $13 million about covered the cost of Society publications. Surely a different scene than existed in Rockefeller Hall in the early 1900's.

Following World War II, the revolution in physics continued, not so much in new fundamental ideas as were discovered in the exciting mid-twenties, but still in surprising ways, extending and largely to be understood by what the European physicists had by then opened up. Exceptions include the developments in quantum electrodynamics, parity non-conservation, CP violation, chromodynamics and fundamental particles, the concept of strings, and the fifth force (?), all modifiers of fundamental concepts which would be reported in the PHYSICAL REVIEW and its *Letters* in its second half century. At this point it should be obvious that by the time of the war, the REVIEW had taken its place—probably first place—among the world's leading journals of physics and that its pages were recording and leading the way for a large part of the subsequent developments in physics.

Many Nobel prizes have been awarded American physicists (and others—chemists and non-Americans) whose prize-winning work was published in the REVIEW. Before war's end we had had Michelson (whose work, however was never described in the journal pages in other than as APS meeting abstracts), Millikan, Compton, Richardson, Langmuir, Anderson, Davisson, Lawrence, and Rabi (more or less in the order of their recognition); Bridgman would be recognized a year after the war. The major work of all of them appeared in the REVIEW. Prize winners of other nations who used the journal at one time or another as we have seen, although not for their award winning work, and in some cases very briefly, include, Bohr, W. H. Bragg, Einstein, Fermi, Franck, Hess, Pauli, Raman, Rutherford, Schrödinger, and Stern; Pauli received his the year of war's end. Beyond all these, we have seen communications—letters and/or contributed articles—of many who would be recognized by the Swedish Academy in years following the cessation of hostilities. There is no doubt that by the beginning of the war our journal had arrived, that it was being widely read, even in Europe. Since World War II, the number of Americans awarded the coveted prize has greatly increased; by 1987 a total of nearly fifty had made the trip to Stockholm after 1945. Out of the some forty physics awards made by the Nobel Committee since that time, Americans have shared or gone solo in twenty-seven of them (as of 1988), their work reported in PHYSICAL REVIEW pages. (In reporting their work in what follows, names of recipients of the award in the post-war years will be noted in italics.) Almost without exception they published their award winning work in our journal. Indeed, some of the noteworthy and award winning European discoveries also appeared in the PHYSICAL REVIEW or PHYSICAL REVIEW *Letters*, for example, the first experimental results on surface topography with the tunneling microscope of *Binnig, Rohrer*, Gerbel, and Weibel in PHYSICAL REVIEW *Letters* (Volume 49, 1982), the application of the quantum Hall effect to the high accuracy determination of the fine structure constant by *von Klitzing*, Dorda, and Pepper in the REVIEW *Letters* (Volume 45, 1980).

With the war over, the physics which perforce had been relegated by the war to the back burner wasted little time in taking off and dazzling the community, some of it facilitated by wartime hardware developments. We think particularly of the elegant and great *Lamb*-Retherford experiment at Columbia, "Fine Structure of Hydrogen by a Microwave Method" (REVIEW, Volume 72, 1947), which confirmed the earlier speculation regarding the non-coincidence of the $^2P_{1/2}$ and the $^2S_{1/2}$ levels in the hydrogen $n=2$ configuration, the outcome (1059 mc shift) of which confirmation opened the way for quantum electrodynamics and the explanation of the shift by *Bethe* in the PHYSICAL REVIEW (Volume 72, 1947), the contributions of *Feynman* (REVIEW Volumes 76 and 80, 1948 and 1950), *Schwinger* and *Tomonaga*, and synthesis of Dyson (Volumes 74, 75, and 76 of the REVIEW, 1948 and 1949). *Tomonaga's* Letter (Volume 74) summarizing the Japanese work in the area is followed with comment by Oppenheimer suggesting that their difficulties result from "insufficiently cautious treatment" of the light quantum self energy. Anyway, Williams and Houston had it right; so did Pasternack. Concerned here also were two other Columbia works: Nafe and Nelson, working with Rabi's beams on the ground state hydrogen (and deuterium) hyperfine levels' separation (the famous 1420 mc transition), turned up a discrepancy (PHYSICAL REVIEW, Volume 73, 1948) which tied in nicely with *Kusch's* (with Foley) electronic g-factor, showing departure from exactly two in Volume 74 (1948) of the REVIEW. In 1958 (Volume 109 of the REVIEW), *Dehmelt* in a variation of his optical pumping work, determined the ratio of the ground state sodium g-factor to the free electron spin g-factor. The technique was exploited by Anderson, Pipkin, and Baird to determine the above hyperfine splittings for all the "hydrogens," H, D, and T, in the PHYSICAL REVIEW, Volume 120, 1960. Some of this atomic physics is very pretty. Subsequently, the difference g – 2 was measured directly at Michigan by Wilkinson and Crane in another beautiful experiment covered in the REVIEW (Volume 130, 1963), following previous but less definitive work (also reported in our journal).

Earliest of the great post-war discoveries published in the PHYSICAL REVIEW, was that of *Purcell*, Torrey, and Pound, followed hard on its heels by that of *Bloch*, Hansen, and Packard in Volumes 69 and 70 (1946) respectively, of nuclear magnetic resonance and nuclear induction (respectively), which have had many ramifications in physics, chemistry, biology, and indeed, medicine. (Medicine is always good to have a ramification in.) *Townes* was shortly working and reporting in the journal on the inversion spectrum of ammonia, following Cleeton and Williams (way back when), and going on to devise and build his ammonia maser (REVIEW Volume 99, 1955) with Gordon and Zeiger, separating the two inversion states by passing the ammonia, as a molecular beam, through a quadrupole electric field. Within a year *Bloembergen* was proposing in REVIEW Volume 104 (1956) his important "Proposal for a New Type of Solid State Maser," based on obtaining a population difference in one pair of levels of a multiple energy level system through saturation of another transition in the multiplet, following similar effect in Overhauser's "Polarization of Nuclei in Metals," reported in Volume 92 (1953) of the REVIEW. Perhaps the ultimate in masers was *Ramsey's* hydrogen maser [PHYSICAL REVIEW *Lett.*, 5, (1960); *Phys. Rev. A*, 138, (1965)], leading to a short time stability of a part or so in 10^{15}, a reasonably precise time piece. Over a long time period, somewhat less stable, but adequate to become world time standard, was the atomic beam cesium clock, capitalizing on Rabi's molecular beam methodology, modified for higher resolution by *Ramsey's* important separated oscillating fields (REVIEW Volume 78, 1950). With such precise timing available, correlation (*a la* Brown-Twiss) of radio noise from distant point sources as received by antennae even continents apart is now routine (sort of) in astronomy's Very Long Base Line Interferometry; and less importantly, one can check relativity's "twin paradox" by flying clocks oppositely around the world, as was done in 1972 (*Science*), wherein the final disparity between fixed and moving clocks was of order 100 ns! Not bad.

Relativity itself continues to be of concern and seems still to be in need of checking every which way to lend assurance of its veracity. (Take a look in *Science Abstracts* Indices under Relativity, both

General and Special.) Pound and Rebka, applying the effect that *Mossbauer* had discovered (1961) in the phenomenally narrow line width of recoilless gamma ray emission of Fe^{57}, measured in the Jefferson Laboratory tower the gravitational red shift of general relativity; the known transverse Doppler shift (of Ives and Stillwell) was also shown in the frequency shift they observed in the difference between hot and cold iron absorber, both communications in PHYSICAL REVIEW *Letters* (Volume 4, 1960). Weber, looking at the vibrational excitation of a large aluminum cylinder, was trying to detect gravitational radiation (Volume 14 of *Phys. Rev. D*, 1976) and set off a flurry of activity (world wide) on reporting [*Phys. Rev. Lett.* 22, (1969)] a positive result, no longer believed tenable. But the technology advances; other higher Q detectors, elaborate and large interferometric detectors are in the works and the search is on for such radiation. (*Taylor*, Fowler, and McCulloch in *Nature*, 277, 1979, believe they have seen the effect of gravitational radiation in the slowing down of the binary pulsar system discovered by *Hulse* and *Taylor* in 1974. Others, including the Nobel committee in 1993, also now believe it.) Shapiro *et al.*, in measuring the time delay in radar echoes from Mercury and Venus observe Einstein's effect of the solar gravitational field on electromagnetic radiation [*Phys. Rev. Lett.*, 22, (1971)]. At higher than earlier sensitivities, the Michelson-Morley experiment has been repeated by *Townes*, Javan, *et al.*, with modern precision in the beating of two coherent sources provided by two infra-red lasers, one in each arm of the "interferometer" [*Phys. Rev. A*, 133, (1964)]. In 1979 this was bettered by a factor of 4000 in a somewhat similar laser experiment by Brillet and Hall (*Phys. Rev. Lett.*, Volume 42) giving an "ether" drift some 10^{-7} of the classical expectation. No surprise in the more accurate results. What was a surprise, however, at this late date was the recognition made by Terrill (PHYSICAL REVIEW, 116, 1959) that the appearance of a box, for example, traveling past an observer at velocities comparable to c, was not that of the box simply contracted in length, but was rather that of the box merely rotated. Provocatively, his article is titled "Invisibility of the Lorentz Contraction."

With contrived population inversion having been shown by *Townes* to make possible his maser oscillator and amplifier, it was not long before the concept was applied to light generation, and the laser came into being. Maiman's ruby laser was first to succeed as a coherent light source (pulsed), briefly announced in *Nature* but described in detail in the PHYSICAL REVIEW, 1961, Volume 123. The continuous output coherent source soon appeared, making its debut in the helium-neon laser of Javan, Bennett, and Herriot [*Phys. Rev. Lett.* **6**, (1961)]. Since then a wide variety of such devices has come about, including, importantly, the CO_2 laser of Patel [*Phys. Rev. A*, **136**, (1964)] and IBM's tunable dye laser; solid state lasers are being widely applied in communications and in the recording industry. In science the laser is everywhere: holography, precision mensuration, spectroscopy, medicine, meteorology, fluids. Of these, our PHYSICAL REVIEW and its *Letters* have especially reported on applications of the tunable dye laser to high resolution spectroscopy. Hansch, *Schawlow* and others at Stanford stand out in this, eliminating the Doppler line broadening in the use of two counter propagating beams through the sample, both beams split from the same laser, one to excite the atomic state of the species being investigated (a molecular iodine line in *Phys. Rev. Lett.* **26**, 1971, and the sodium D lines in the *Letters* also of Volume 26), the other, a monitored weak beam, to check the degree of saturation produced by the first as evidence that excitation has taken place [*Phys. Rev. Lett.* **27**, (1971)]; only those atoms of zero longitudinal velocity show up (an effect *Lamb* showed in his "Theory of the Optical Maser" [*Phys. Rev. A*, **134**, (1964)] to be responsible for what has come to be known as the Lamb dip at the peak of the "lasing" line profiles). In this beautiful way, the fine structure of the Balmer alpha line was completely resolved, with the Lamb shift obvious, a memorable contribution to *Nature* in 1972 (Phys. Sci. Sect. 235). Doppler broadening has also been eliminated and some advantage accrues with two photon excitation from two half energy, counter propagating beams; using this technique, Hansch *et al.* studied the 1s to 2s (Lyman alpha) transition in hydrogen and obtained the Lamb shift for the ground state [*Phys. Rev. Lett.* **34**, (1975); PHYSICAL REVIEW A, **22**, (1980)]. (In a molecular beam experiment, utilizing Ramsey's separated oscillating fields, Fabjan and Pipkin had already measured the shift for the $n=3$ level, reported in the PHYSICAL REVIEW A, **6**, 1982.) In a modification of their original saturation experiment, the Stanford group made a factor of ten improvement in a measurement of the Rydberg constant, reported to *Phys. Rev. Lett.*, **32**, 1974. With a point contact mixer diode, it has become possible (and done) literally to count (in a series of harmonic generators deriving from an oscillator of lower, known, countable frequency) the oscillations per second made by an infra-red He-Ne laser, leading to a precision determination of the all important constant, c [*Phys. Rev. Lett.*, **29**, (1972)] (Barger, Hall, Evensan *et al.*). As a result, the standard meter is now defined in terms of c, and the distance between two scratches on a certain platinum–iridium bar in Paris has become irrelevant. Modest frequency light beats between two lasers is today no great trick, but light beating was actually demonstrated well before the laser era in an important and difficult experiment proposed in a letter (Volume 72) to the REVIEW from Forrester, Parkins and Gerjuoy in 1947, wherein beats between the Zeeman components of ordinary incoherent light would be mixed at a photo surface, the electrons emitted from which would excite a microwave cavity. Forrester, Gudmundsen, and Johnson reported on the successful completion of the experiment eight years later in the PHYSICAL REVIEW (Volume 99, 1955).

Precision spectroscopy of another sort developed elsewhere. Following in his father's high resolution X-ray spectroscopy, and on the basis of magnetic analysis of electrons ejected by X-rays, *Kai Siegbahn* and his group in Uppsala were doing precision photo-electron spectroscopy, mostly reported in Sweden. In a very short letter to the Review (Volume 110, 1958) Sokolowski, Nordling and *Siegbahn* reported the "Chemical Shift Effect in Inner Electronic Levels of Cu Due to Oxidation," the full details to be published in the Swedish *Arkiv. Fysik*. (Another instance of father–son Nobelists: we have the Thompsons, the Braggs, the Bohrs; and there is also the brothers-in-law pair—Townes and Schawlow.)

Lasers are playing an important role in particle trapping, which has recently taken on importance and in which experimentalists have shown amazing ingenuity. Neutral particles have been caught in a "viscous" photon sea created by counter propagating laser beams on three axes (Chu *et al.*, *Phys. Rev. Lett.* **55**, 1985), the radiation confining the atoms, both cooling them and holding them for a short spell until they wander out of the trap, like flies from a pot of molasses. Ashkin has described in the *Letters* (Volume 40, 1978) a trap wherein atoms are confined by two counter propagating, *focused* laser beams near (slightly below) the atoms' resonant radiation, forces arising from gradients in the radiation field. Or atoms can be trapped at a magnetic field minimum (through action on the magnetic moment), cooled first by successive momentum transfers upon successive absorptions and re-emissions of resonant laser light, very clever means devised for allowing the light to match the reduction in Doppler shift as the particles cool, described in one method by Pritchard *et al.* in a note from MIT to PHYSICAL REVIEW *Letters*

(Volume 58, 1987). The MIT group has trapped thusly a swarm of sodium atoms cooled to °mK, separating, unbelievably, different hyperfine states at different magnetic field positions in the earth's gravitational field. Another method comes in a collaborative effort of the Chu and Pritchard groups. Atoms are trapped by the scattering of near resonant light; a very weak magnetic field differentially shifts the atomic magnetic sublevels across the trap so atoms preferentially scatter photons which nudge them back to the trap center, cooling them also to a fraction of a °mK, described in the REVIEW Letters (Volume 59, 1987). Amazing business. (A nice survey article on trapping, with references: *Science*, 1988, February 19.)

Charged particles have also been trapped, somewhat more easily, in Penning traps; tricks are played to reduce the number to a single ion, at which point some remarkable physics is done, much of it at the University of Washington. Wineland, Ekstrom, and *Dehmelt*, at the University, have trapped a single electron for a period of weeks—a sort of giant atom that has been labeled "geonium" [*Phys. Rev. Lett.*, **31**, (1973)] on which to do fundamental physics. In such a trap, Gabrielse, *Dehmelt*, and Kells have observed [*Phys. Rev. Lett.*, **54**, (1985)] a relativistic hysteresis in the cyclotron motion of a single trapped electron as the drive frequency is decreased, an effect predicted by Kaplan in Volume 48 of REVIEW *Letters*, 1982. It is not necessary to go to a 184" cyclotron to run afoul of Bethe and Rose's "limit" to the cyclotron; in Gabrielse's trap with an electron energy of less than an ev, the change in electron mass is a part per million, more than enough to show the pronounced effect. Van Dyck, Schwinberg, and *Dehmelt* reported in the REVIEW *Letters* (Volume 38, 1987) the determination of the empirical ratio of the spin flip to cyclotron frequencies as 1.001 159 652 410 (200), "the most precisely known characteristic of an elementary particle," it is modestly claimed. However, we would not bet on that claim today, and certainly not tomorrow. (For a nice discussion of this experiment deriving the g anomaly, see the article of Ekstrom and Wineland in the August 1980 issue of *Scientific American*.)

In another *Letters* (Volume 47, 1981), the ratio of the g-factor for the positron to that for the electron was found to be $1 + [(22 \pm 64) \times 10^{-12}]$. Using a similar trap, one very precisely characterized and known, Williams and Olsen at the Bureau of Standards described a new measurement of the proton gyromagnetic ratio [*Phys. Rev. Lett.*, **42**, (1979)], leading to a new derived value of the fine structure constant good to a part in 10^7. That would give the theorists something to chew on, as exactly that is what they have been doing. Quantum effects can be studied; Bohr's quite discontinuous jumps between quantum states of an atomic system, in an experiment suggested by Cook and Kimble [*Phys. Rev. Lett.*, **54**, (1985)], have been observed, for example, by the Washington group in REVIEW *Letters* (Volume 56, 1986) for a single "cold" barium positive ion. The signature of one such ion was shown as a bright spot in a photograph of *Dehmelt's* reproduced in a PHYSICAL REVIEW A paper (Volume 22, 1980). Inhibited spontaneous emission of a trapped electron has been observed by the Seattle-ites and reported by them in the *Letters* (Volume 55, 1985). We are really getting down there in all this. At CERN, an international and large group reported [*Phys. Rev. Lett.*, **57**, (1986)] anti-protons at 3 keV trapped and held for periods as long as ten minutes, with as few as five of the particles remaining at the end. There is talk of trapping anti-hydrogen atoms to get hold of other fundamental behavior. Incredible. On a different level, fundamental and going back to acoustics and the speed of sound, was the determination of the universal gas constant to less than two parts per million by a Bureau of Standards and University of Delaware collaboration in 1988 (*Phys. Rev. Lett.*, **60**).

Of all the work reported in the PHYSICAL REVIEW during its century of life, that which has most affected the world generally is probably the small, bench-top discovery by *Bardeen* and *Brattain* at Bell Laboratories of the transistor, leading as it has almost to the whole of solid state electronics. Where the all pervasive computer, which it has made possible, will lead us, no one can foretell. That revolution is still upon us. The announcement of the discovery came in three letters in the PHYSICAL REVIEW (Volume 74, 1948), including one by *Shockley* and Pearson, the former directing the group effort. Prior announcement was made at a conference for the daily press, the long gone *New York Herald Tribune* giving it nice and substantial coverage, and the *New York Times* alloting it a two- or three-inch item back on the entertainment page. A decade later, after the point contact transistor had given way to Shockley's grown junction transistor, *Leo Esaki*, in Japan, wrote in Volume 109 (1958) of the REVIEW on a "New Phenomenon in Narrow Germanium p-n Junctions." He might have added that it was heavily doped material. In his diode, electron tunneling through the thin barrier gave the i–V characteristic a *negative* slope over a certain range. Work in germanium and silicon (shades of Miss Wick) would be hot and heavy for years in many laboratories, both industrial and academic, here and abroad, results duly reported in our journal. (Not only has there been keen competition between solid state physicists but it is also present in a number of the sub-fields of our discipline—for example, high energy physics—to the degree that notable discoveries are frequently fed to the press, as in this instance, before the work appears in scientific journals, the REVIEW or otherwise. It is not a new ploy.) In a talk at the APS Symposium on the occasion of the 25th anniversary of the transistor discovery, J. B. Fisk, Bell Labs President, told of its being a non-controversial, benign development—no pollution there. In private conversation afterwards, *Brattain* allowed as how they had not counted on noise pollution!

Of the fundamental discoveries beyond those in quantum electrodynamics, undoubtedly the most important was the solution to the so-called τ-θ puzzle, two heavy mesons, τ and θ, having the same mass and lifetimes. How could it be? Dalitz discussed this in a REVIEW article in Volume 94 (1954), which was important to the conjecture made by *Lee* and *Yang* (in REVIEW Volume 104, 1956) that τ^+ and θ^+ were two different decay modes of the same particle, which process might not strictly conserve parity, a suggestion solidly confirmed by a Letter to the REVIEW (Volume 105, 1957) in the beautiful experiment of Madame C. S. Wu of Columbia, working with Ambler and a group at the Bureau of Standards. Before the first publication, the news got around, prompting another Columbia group (Garwin, *Lederman*, and Weinrich) to study the decay of positive muons made in their 85 MeV pion beam. The parity non-conservation showed up clearly, and the magnetic moment of the free muon was also determined, published in a Letter immediately following that of the Wu disclosure. There could have been a third communication; Friedman and Telegdi, using emulsions in Chicago also observed the asymmetry in the emitted electrons, but their Letter was bounced and only appeared in the next issue. Telegdi was pretty mad. But it was Wu *et al.* in priority anyway and clearly parity was not conserved. The non-conservation in atomic physics was detected thirty years later (*Phys. Rev. Lett.*, **55**, 1985) in a crossed atomic cesium beam-laser beam experiment by Gilbert *et al.* That parity might not alway be conserved was raised in a suggestion made by *Purcell* and *Ramsey* in a REVIEW Letter in 1950 (Volume 78), proposing to do NMR on the neutron in an electric field. Subsequent to all this, *Feynman* and *Gell-Mann* at Cal Tech wrote an important paper on the "Theory of the Fermi Interaction" in REVIEW Volume 109 (1958), a "unique, weak, four fermion coupling—equivalent to equal amounts of vector and axial vector coupling with two component neutrinos and conservation of leptons," to quote their abstract. Similarly fundamental like the parity non-conservation, but in no way as shocking (less also to be understood by non-particle physicists), were the reports in PHYSICAL REVIEW *Letters* of *Fitch* and *Cronin* on the simultaneous violation of charge conjugation and parity inversion symmetries (Volumes 13 and 15, 1964 and 1965). Shocking indeed, however, would be confirmation of a fifth force proposed by Fischbach *et al.* in their PHYSICAL REVIEW *Letters* of Volume 56, 1986, "Re-analysis of Eotvos' Experiment," in which it is suggested that variations found in the experiment seem to depend on the baryon number per unit mass of the sample nuclei and could be understood by postulating the existence of a new force of intermediate range. Many experiments are already in progress to check the hypothesis but as reported in various REVIEW *Letters*, the results are divergent and inconclusive, one group even finding a sixth force!

Nuclear physics, while active, was gradually yielding place to high energy physics and fundamental particles. An exception was the work at Cal Tech's Kellogg Laboratory where *W. A. Fowler* and the Lauritsons decided to continue in "classical" low energy physics. The 1950 REVIEW index, to illustrate, is replete with references to their (and collaborators') work. *Fowler's* aim was to take such data and apply it to reaction rates in stars; it resulted in not only his Nobel prize with *Chandrasekhar* in 1983, but in the famous "B^{2}FH" review paper with Hoyle and the two Burbridges, "Synthesis of Elements in the Stars" (*Rev. Mod. Phys.*, 1957). In this realm too is a well-known Letter to the REVIEW (Volume 73, 1949) on the "Origin of the Chemical Elements," noted less for its importance than for its authors, Alpher, *Bethe*, and Gamow, a sequence certainly put together by the last named. Notable progress in nuclear physics came largely in the area of nuclear structure. In 1948, *Aage Bohr* and Weisskopf had a paper in Volume 74 of the REVIEW on the "Influence of Nuclear Structure on the Hyperfine Structure of Heavy Elements," following up *Aage Bohr's* suggestion two years earlier in Volume 73 that internal structure of the deuteron might explain the theoretical discrepancy shown in the hyperfine structure of H^2. *Maria Goeppert-Mayer* contributed her important nuclear shell model with its "magic numbers" first in a Letter and paper in Volume 78 of 1950. *Rainwater*, with many REVIEW papers in time of flight, slow neutron spectroscopy to his credit, disclosed something different in the journal (Volume 79, 1950) with his spheriodal nucleus, which could explain large quadrupole moments not expected in the spherical shell model. In 1951, Martin Deutsch from MIT reported in a REVIEW Letter of Volume 82, "Evidence for Formation of Positronium in Gases," following it up with Letters in succeeding volumes. Muonium would not be far behind. First evidence, indirect, came in the γ-emission entailed in the absorption of μ- in H^2. by Panovsky, Aamodt, and York [*Phys. Rev.*, **78**, (1950)] considered earlier in the REVIEW by Fermi and Teller ("The Capture of Negative Mesons in Matter") in 1947 (Volume 72). The first direct evidence for the entity came in 1952 (*Phys. Rev.*, **88**) from Rochester with "X-rays from Mesic Atoms" by Camac, McGuire, Platt, and Schulte. "Studies of X-rays from Mu-mesonic Atoms" were carried out subsequently by *Rainwater* and *Fitch* in REVIEW Volume 92, 1953. More remarkable perhaps is Vernon Hughes "Formation of Muonium and Observation of its Larmor Precession" in Volume 5, 1960 of PHYSICAL REVIEW *Letters*. In collaboration with Marder and Wu, in REVIEW Volume 106 (1957), he had measured the "Hyperfine Structure of Positronium in its Ground State," improving the measurement of Deutsch and Brown five years earlier in Volume 85. There is no doubt about the existence, however brief, of these exotic species; the Index entries in *Science Abstracts* for each of the rare atoms is today extensive. The agreement of experiment with theory, which presumably is not encumbered by internal structure of the involved particles of each species, is good. The theory involves the important so-called *Bethe-*Salpeter equation, frequent reference to which occurs throughout atomic physics and which is developed in their consecutive REVIEW papers of Volume 84, 1953, ending with "Relativistic Equation for Bound State Problems." *Willard Libby* had a noteworthy Letter in REVIEW Volume 69 (1946), of as much importance to archeology as to physics, with his "Atmospheric Helium Three and Radiocarbon from Cosmic Rays." Regarding cosmic rays, *Alfven* was still worrying over the source with his "On the Solar Origin of Cosmic Rays" in Volumes 75 and 77 (1949 and 1950) of the REVIEW. But in a better known paper, also in Volume 75, Fermi with "On the Origin of Cosmic Radiation"argues that the particles come from interstellar space, accelerated by collision against moving, random, magnetic fields.

In 1947, continuing work they had started *sub rosa* in war torn Italy, Conversi, Pancini, and Piccioni reported in a REVIEW Letter (Volume 71) on some strange results they had observed in the decay of cosmic ray mesons in graphite, aluminum, and iron. In the same volume, 100 pages later, Fermi, Teller, and Weisskopf write of the Italians' letter that the result "seems to indicate that the interaction of mesotrons with nucleii according to the present schemes is many orders of magnitude weaker than usually assumed." Harmony could be restored by postulating another, short lived meson, from the decay of which came the one then known, as Marshak and *Bethe* discussed in REVIEW Volume 72, "On the Two Meson Hypothesis" of Marshak. Later in the year, 1947, *Powell et al.* in *Nature* (Volume 159) reported the discovery of that short lived particle, *the* meson, the pion, in some striking photographic emulsion tracks exposed to cosmic rays at the top of Pic du Midi. It would not be long before pions would be produced routinely in large accelerators.

The ghostly and elusive neutrino of beta decay was detected at a nuclear pile by Reines and Cowan, and reported in a Letter to the REVIEW (Volume 92, 1953), opening the way for neutrino physics, and in fact, to neutrino astronomy in surprising ways fifteen and thirty-five years ahead. Despite the difficulty and low counting rates met in working with them, it has become feasible to do high energy neutrino experiments with beams of neutrinos created in the decay of pion beams which are "manufactured" in the bombardment of low Z materials with high energy protons. This was a proposal made by Schwartz in *Phys. Rev. Lett.*, **4**, 1957, as a technique likely to be available when sufficiently high energy proton beams came on line. Five years later (REVIEW *Letters*, **9**, 1962), with such a neutrino beam at the Brookhaven "Cosmotron" proton accelerator, a Brookhaven–Columbia consortium (including *Schwartz, Lederman,* and *Steinberger*) showed convincingly in an eight-page landmark *Letters* paper that there had actually to be two neutrinos, one for the electron and one for the muon, suggested in an analysis by Feinberg in a Letter to the PHYSICAL REVIEW in 1958 (Volume 110).

Particles were discovered at a great rate as the big post-war machines came into operation, the first of which was the 184" Berkeley cyclotron (*Phys. Rev.*, **71**, 1947, the machine frequency modulated to get around relativity), and which was essentially waiting in the wings throughout the war. The π^- meson was seen there in emulsions exposed in the bombardment of carbon by alpha particles accelerated to 380 MeV by the machine (Gardiner and Lattes, *Science*, 1948), and shortly thereafter the REVIEW carried their report of similar observation of the π^+ meson (Volume 75, 1949). A lifetime measurement of the π^- was soon done in a nice experiment at the machine by Richardson in a REVIEW paper of Volume 74, 1948, soon after the report of its production. This was followed a year later with a π^+ lifetime measurement at the same site by Martinelli and Panofsky in REVIEW Volume 77, 1949. Evidence for the π^0 came from Rochester also in 1949, Volume 76, in an emulsion display of the "R(ochester) Star" captured by Kaplon, Peters, and Bradt in a balloon at 100,000 feet. Their Letter on the energetic event was joined immediately by another Rochester Letter (Marshak) discussing implications of the "R Star"; a neutral π existing for 10^{-15} sec seemed involved.

High energy accelerators, be it cyclotrons, synchrotrons, synchrocyclotrons, linear accelerators, storage rings, colliders, what not, proliferated well beyond the confines of Berkeley. Besides their making possible the production of a plethora of various particles, announced sporadically in the REVIEW, its *Letters*, or the daily press, those machines around which electrons orbited had another and side benefit. In 1947, Volume 71 of the PHYSICAL REVIEW had a Letter telling of the observation of visible light from electrons radiating in the 70 MeV synchrotron under development at G.E. (confirming *McMillan's* and *Vecksler's* idea), directed radiation, the spectral distribution of which over the visible was analyzed there by D. Langmuir, Elder, and Pollock in Volume 74, 1948. This synchrotron radiation, not so named at the time, was considered by Iwanenko and Pomeranchuk (1944) and *McMillan* (1945) as we have noted, and much earlier by Schott, and unknowingly previously observed in such astronomical objects as the Crab nebula. Here on earth, it has been both boon and bane to nonlinear accelerators, representing both a major energy loss for the machine and a useful by-product. In 1953 Corson measured (REVIEW Volume 90) the energy loss experienced by electrons in the Cornell 300 MeV synchrotron. Others there demonstrated (PHYSICAL REVIEW, **102**, 1956) its utility by showing the copious presence of the radiation down even below 110Å, with a striking continuum exposure and the beryllium K-edge at 100Å delineated, their analysis of the spectral and beam-like, angular distribution confirming the elaborate theory of *Schwinger* [*Phys. Rev.*, **75**, (1949)]; Corson's loss measurement was also in agreement. The first new physics done with the radiation was that of Madden and Coddling at the Bureau of Standards 180 MeV machine, reported in the REVIEW *Letters* (Volumes 10 and 12, 1963 and 1964). The radiation has since come to be widely used even into the high energy X-ray regime and a large enterprise around the world has developed exploiting it, several dedicated machines for its production both now on line and in the planning. It is particularly useful and easier to come by in electron storage rings, concept of which was proposed by G. K. O'Neil in a Letter to the PHYSICAL REVIEW, **102**, 1956.

Helpful in the search for "fundamental" particles was the bubble chamber of *Donald Glaser*, grown from his thimble-sized flask of ether that he reported first in a Letter to the PHYSICAL REVIEW (Volume 91, 1955) to monsters six feet in diameter filled with liquid hydrogen, the subsequent developments very much due to the lead, boldness, and successes of *Luis Alvarez*. With his intermediate, fifteen inch model he and his group investigated and found resonances in the pion systems, $\lambda - \pi$, $\kappa - \pi$, and $\xi - \pi$, produced in the collisions of K^- mesons on hydrogen, reported in the REVIEW *Letters*, Volumes 5 and 6, 1960 and 1961. World wide, millions of beautiful particle track displays in bubble chambers have been photographed, significant ones reproduced in scientific journals, including the PHYSICAL REVIEW, as evidence of this or that new exotic particle species. And plenty were found, so numerous as to bewilder many a high energy physicist. One such not too bewildered was *Murray Gell-Mann*, who brought some order to the multitude in his "eight-fold way," described in his paper "Symmetries of Baryons and Mesons" (REVIEW Volume 125, 1962) to be followed two years later with his invention of quarks as particle constituents (Zweig independently cooked up a similar scheme); the Dutch journal *Physics Letters* published this landmark suggestion. Crowning the "eight-fold way" was the identification in a bubble chamber event by a Brookhaven, 31 man team of the one remaining missing member of the Gell-Mann ordering. In 1964 the REVIEW *Letters* (Volume 12) carried the communication: "Observation of a Hyperon with a Strangeness Minus Three"—the Omega minus, with mass and spin in agreement with what the theory predicted. Very nice. In 1955, *Chamberlain*, *Segre*, Wiegand, and Ypsilantis, from Berkeley, reported in the PHYSICAL REVIEW (Volume 100, 1955) the discovery of the antiproton; across the Bay, at Stanford, *Hofstedter* was studying the elastic ("coherent") scattering of high energy electrons, finding, in a Letter to the REVIEW (Volume 98, 1955), the proton to have a radius of 10^{-13} cm and the Coulomb force to be OK at distances down by a factor of ten less than that. More significant and striking perhaps were the results of a collaboration (for example, in two adjacent Letters [*Phys. Rev. Lett.*, **23**, (1969)] from SLAC and MIT (1990 Nobelists *R. Taylor, H. Kendall,* and *J. Friedman, et al.*) and a later paper in the PHYSICAL REVIEW D, Volume 5, 1972) on deep inelastic ("incoherent") scattering, showing the proton to have internal structure, indicative of the until then merely speculative partons-quarks. In consequence of the experiment, Bjorken and Paschos in the REVIEW (Volume 185, 1969) had a paper concluding that the first work (at 6° and 10° scattering angle) had shown the quarks (partons) to be essentially free on the inside of the proton. It was titled "Inelastic Electron-Proton and γ–proton Scattering and the Structure of the Nucleon." At Stanford again, and at Brookhaven independently, *Richter* and *Ting* respectively (both backed by significant teams), in a very narrow resonance at 3.1 GeV found the ζ particle and the J, also respectively, each identical to the other and dubbed "charmonium," for the "charmed" quarks of which it is presumed to be constituted. Very shortly thereafter, Stanford found another resonance at 3.7 GeV. The initial discoveries were reported in three communications (one also from Frascati, Italy, where they had heard the news), back to back, in *Phys. Rev. Lett.*, **33**, 1974. The *Letters* even had an editorial commenting on the simultaneous announcement from three laboratories and congratulating the teams for the momentous discovery. The "charmed" quark was an entity introduced by Bjorken and *Glashow* (in *Physics Letters*) and used by *Glashow*, Iliopoulos, and Maiani in their PHYSICAL REVIEW D paper (Volume 2, 1970), "Weak Interactions with Lepton–Hadron Symmetry," a model of weak interactions in which a charmed, massive, vector-boson interacts with currents constructed out of four basic quark fields p , η, and λ forming an SU(3) triplet (obviously reading from their description of the model), "and the fourth, p', has the same electric charge as p but differs from the triplet by one unit of a new quantum number c for charm." *Weinberg* had an important role to play in this with his earlier paper, "A Model of Leptons" (*Phys. Rev. Lett.*, **19**, 1967), rather crucial to the joining of the weak force to the electromagnetic. It had been preceded by a paper with similar treatment by *Abdus Salam* and J. C. Ward in the PHYSICAL REVIEW (Volume 136B, 1964), "The Gauge Theory of Elementary Interactions," of which *Weinberg* was unaware. This was preceded in turn by work of N. Cabibbo [*Phys. Rev. Lett.*, **10**, (1963)], "Unitary Symmetry and Leptonic Decays"; we keep hearing of the "Cabibbo Angle." Inherent in the theory were "neutral currents," evidence for which was forthcoming quite later from Fermilab (also from CERN) in a communication from Benvenuti *et al.* [*Phys. Rev. Lett.*, **32**, (1974)] and from Barish *et al.* [*Phys. Rev. Lett.*, **34**, (1975)]. Another paper playing a role was that of Peter Higgs, "Spontaneous Symmetry Breakdown without Massless Bosons," in the PHYSICAL REVIEW, **145,** 1966; one keeps hearing of Higgs fields, Higgs theory, the Higgs boson, Higgs what not; his boson remains to be discovered in the next generation of high energy machines. One also keeps hearing throughout particle theory of *Yang*–Mills this and that. The pair made a major breakthrough, early at that, central to the whole unification business, in their oft referred to PHYSICAL REVIEW paper (Volume 96, 1954) on "Conservation of Isotopic Spin and Isotopic Gauge Invariance." Symmetry and gauge loom large in modern field theories, if not of all physics. In *Phys. Rev. Letters* (Volume 32, 1974) Georgi and *Glashow* conjectured that the gauge group SU(5) brings "Unity of all Elementary Particle Fields;" not necessarily the last word but not just "idle" speculation either, an idea "to be taken seriously." Proton decay is inherent in the theory, which has resulted in a number of large deep-in-the-earth experiments around the world hunting for it. A well-known paper contributing to these developments comes from a surprising quarter; *P. W. Anderson*, earning his Nobel award for work in solid state physics, such as "Localized Magnetic States in Metals" in the REVIEW (Volume 124, 1961), clarifies a point in *Yang*–Mills theory, that their vector boson does not necessarily have zero mass. He considers the case from metals electron plasma physics in his "Plasmons, Gauge Invariance and Mass" in the REVIEW Volume 130, 1963. How all of this gets synthesized into a Grand Unification Theory (GUT), which is the aim, is beyond the purview of this brief selection of papers germane to the dream (not to mention being beyond the comprehension of the writer). Grand unification has long been a goal of physics and it may be on the way. Gravity is still not unified, although today many theorists see "strings" as the answer to everything (including distribution of the galaxies), even with their ten dimensions, six of which are rolled up and "compacted" in some way. Papers on strings abound in the REVIEW and elsewhere. The optimists see them in relation to late 20th century physics as relativity and quantum theory were to the physics of the early part of the century, i.e., revolutionary. We'll see, and the PHYSICAL REVIEW and its progeny will be carrying many of the developments; most of us won't understand in any event, leaving that to the twelve (or more) supermen, like those who alone in all the world were said to understand Einstein back when.

With strings in space, high energy physics and astronomy have come into some juxtaposition, the universe conceived to have begun in an enormous explosion from a singularity, following which, in the next small fraction of a femto-femtosecond or so, strings, all manner of exotic particles, carriers of force fields, and symmetries evolve or devolve. Strengthening the view was the discovery, in the microwave spectral region, of the remnant black body radiation pervading all of space, cooled down from the unimaginably high temperatures at the start. *Penzias* and *Wilson* made this discovery at the Bell Laboratories (birthplace of the field of radio astronomy), reported in the *Astrophysical Journal*. A Berkeley consortium investigating the infra-red portion of the presumed radiation, from 3 to 40 cm^{-1} (REVIEW Letters,Volume **34**, 1975), found the peak of the distribution at about 7 cm^{-1}, confirming the interpretation. Present (1990) temperature: 2.73 °K. The big explosion has problems which Alan Guth in PHYSICAL REVIEW D, Volume 23, 1981, gets around in his "Inflationary Universe: A Possible Solution to the Horizon and Flatness Problems." Not that his scenario has no problems of its own, but "inflation" and "bubbles" are popular aspects of cosmology today. Also in astronomy, the black hole of Oppenheimer and Snyder, we spotted earlier, came to be believed in, and rotating neutron stars (pulsars) were discovered. Gravitational lenses (recall Zwicky; discussed anew by Liebes in the REVIEW, Volume 133 B, 1964) were looked for and seemingly observed, and in 1987 a supernova, SN-1987a, in the large Magellanic cloud, sent neutrino signals to two large buried detectors (searching for proton decay), one in Japan and one beneath Lake Erie (*Phys. Rev. Lett.*, 58, 1987), eliciting all kinds of excitement in both astronomy and physics. Another and even more audacious detector set up in the Homestake gold mine a mile below the Black Hills of South Dakota searched specifically for solar neutrinos. With it, Davis, Harmer, and Hoffman, in Volume 20 of PHYSICAL REVIEW *Letters* (1968) reported detecting only a third or quarter of those that should be found on the basis of current theory of stellar energy production. Further data accumulation over two decades has only confirmed the result, which has been the subject of much worry, starting right off with the *Letters* of Bahcall, Bahcall, and Shaviv, immediately following the first report in 1968. On the basis of a Russian suggestion, *Bethe* in our REVIEW *Letters* (Volume 56, 1986) considered the neutrino oscillations proposed by Pontecorvo, and developed by Wolfenstein in the PHYSICAL REVIEW D (Volume 17, 1978), as quite possibly the answer to the puzzle; in the scenario some of the electron neutrinos are converted to muon neutrinos to which Davis' detector is not sensitive. Large gallium detectors under way (1988) in Russia and in Italy were possibly to verify the supposition.

Back to earth were more mundane matters. Low temperatures in experimental physics became routine with the prevalence of liquid helium. At 4°K, Bowers, Legendy, and Rose reported from Cornell in REVIEW *Letters* (Volume 7, 1961) observation of "helicon" waves in metallic sodium, waves in the electron gas of the metal the analogue of "whistlers" in the ionosphere. The REVIEW itself (Volume 127, 1961) brought the details. At long last, in 1957 in Volume 106 of the REVIEW, *Bardeen, Cooper,* and *Schrieffer* presented in a *Letter* an explanation of superconductivity which has withstood all tests; a long article on their BCS theory followed in Volume 108 (1957); electrons become bosons through pairing arising by phonon interaction as *Bardeen* had shown, on the basis of the important result that T_C in mercury depends on the isotopic mass (Reynolds, Serin, Wright, and Nesbitt in *Phys. Rev.* **78**, 1950 and Maxwell in the same volume), in his REVIEW papers of Volumes 79, 80, and 81 (1950 and 1951). The ideas of electron pairing and a Bose-Einstein condensation developed over a number of years, starting with London's 1935 REVIEW suggestion, cited earlier, from the analogue with helium superfluidity. In 1946 Ogg at Stanford in a Letter to the PHYSICAL REVIEW (Volume 69, 1946) hypothesized a Bose–Einstein condensation of paired electrons trapped in F′ centers (today's polarons?) as an explanation of the superconductivity he reported in his frozen, dilute, sodium-liquid ammonia solutions, existence of which superconductivity, however, is not widely accepted. In 1954 Schafroth in a Letter to the REVIEW (Volume 96, 1954), having shown in an earlier Letter in the same volume that a charged boson gas below its condensation point is a superconductor, showed that if the total interaction energy (Coulomb, lattice vibrations) is such as to produce resonant states of pairs, then superconductivity should set in. In several articles covering nearly sixty pages of one REVIEW issue of 1955 (Volume 100), several authors (Chester, Blatt, Butler, Schafroth) consider matters pertaining to superconductivity and superfluidity. In one of these articles, Schafroth strengthens his assertion concerning the charged boson gas. In 1956 *Cooper* introduced his spacially extended pairs to produce a small energy gap between the ground state and first excited state (REVIEW Volume 104) leading to the BCS picture. The pairs differ from those of Schafroth which are of closely bound, pseudo-molecular entities. The Cooper pairs "are not localized in (that) sense and (the) transition is not analogous to a Bose–

Einstein condensation," to quote B, C, and S in their footnote comment. Pairing has not yet (1988) been explained, but believed requisite, in the case of the high temperature, ceramic oxide superconductor discovered at 40°K by *Bednorz* and *Muller* of IBM (Zurich), which Chu of Texas (Houston) followed up with superconductors of similar ilk going critical at temperatures above that of liquid nitrogen, reported in Volume 58, 1987, of REVIEW *Letters*. Little time was lost in awarding the IBM duo the 1987 Nobel prize; one notes that the prize of the preceding year went to another IBM duo (Zurich again), *Binnig* and *Rohrer*, for their scanning tunneling microscope, with which in *Phys. Rev. Letters* (Volume 49, 1982), they demonstrate, atom by atom, the undulating structure of a gold surface. The low temperature developments elicited vast interest at a wild session at the solid state APS meeting in New York in the Spring of 1987, not to mention many low temperature physics laboratories and theory centers. Also at low temperatures, in the post-1945 years, Josephson theorized in the new European journal *Physics Letters* (Volume 1; PHYSICAL REVIEW *Letters* could not have all the good stuff) on electron tunneling between two superconducting metals separated by a thin film insulator. In 1960, *Giaever* at G.E. investigated the energy gap and density of states in a superconductor, using an Al-thin Al_2O_3-Pb tunneling sandwich, contrasting the behavior with the lead side normal as opposed to it at a lower temperature as a superconductor (*Phys. Rev. Lett.*, **5**). The results were in remarkable accord with the BCS theory of superconductivity. Behavior of the Josephson junction to both DC and AC voltages (*Anderson* and Rowell, *Phys. Rev. Lett.* **10**, 1963, and Shapiro, *Phys. Rev. Lett.*, **11**, 1963, respectively) has surprising applications, not the least of which is the simple involvement of fundamental physical constants; with an rf voltage of frequency v, a *V versus I* plot shows flat steps in the characteristic which are separated by a voltage of $hv/2e$, the factor two indicative of the paired electrons of the BCS theory. Also involving of fundamental constants in a simple way is the quantum Hall effect discovered by *von Klitzing*, as he pointed out in a PHYSICAL REVIEW *Letters* communication (Volume 45, 1980). The two-dimensional electron gas in his semiconductor sandwich makes possible the determination of alpha, the fine structure constant, to high accuracy; with good voltage measurements (not easy) the AC Josephson effect yields a good value for h/e (or given the frequency, v, the value $hv/2e$ serves to define the volt). Theorists are becoming hard pressed to match the number of decimal places the experimen-

talists in various areas are now presenting them. But they are working at it. At Cornell, Kinoshita for example, with Cvitanovic, in 1972
calculated the electron moment anomaly, $(g - 2)/2$, to the sixth significant figure ($\times 10^{-9}$) using 72 Feynman diagrams of order $(\alpha/\pi)^3$
(*Phys. Rev. Lett.*, 29), finding agreement in the theoretical and experimental values of the anomaly to a part or so in 10^6. The calculation
has by today been further refined, utilizing 891 (!) of Feynman's
ubiquitous diagrams for increased (about 100 times) accuracy, presumably to be reported ultimately in PHYSICAL REVIEW pages. In the
matter of derived physical constants, experimentalists and theorists
are taxing each other's ingenuities, not to mention perseverances.

Two Josephson junctions arranged in parallel show quantum
interference phenomena, reported by a group under Mercereau at
the Ford Motors Laboratory in two communications to *Phys. Rev.
Letters* (Volume 12, 1964). The whole cold business had led to the
SQUID, a super-sensitive detector of magnetic fields and, in the year
before Josephson's work, to the discovery of flux quanta, predicted
in 1950 by London. Independently, the phenomenon was found by
Deaver and Fairbank and by Doll and Nabauer in Germany, both
reported in the same REVIEW *Letters* (Volume 7, 1961) in company
with theoretical notes by *Onsager* and by Byers and *Yang*. The quantization of flux was shown later in even an ordinary cold, gold ring
(not so ordinary, in fact; it was less than half a micron in diameter
and of thickness less than 10% of that) by an IBM group [Webb *et al.*,
Phys. Rev. Lett., **54**, 1985)], every change of flux of h/e through the
ring showing one oscillation change in current across the ring, up to
as large a magnetic field as the experiment was carried.

Quantum interference was actually predicted earlier in a rather
unexpected quarter. While the magnetic vector potential was known
from the beginning of quantum mechanics to be involved in the
phase of the wave function, it was not until 1959 in Volume 115 of
the PHYSICAL REVIEW that Bohm and Ahranov pointed out in an
article that if an electron beam were divided and sent on either side
of a long solenoid *confining* a magnetic field, that nevertheless, the
electrons on the outside would know of it and would show it in their
field sensitive interference pattern on the far side of the solenoid.
Within a year, Chambers, at Bristol, had demonstrated the reality of
the prediction in *Phys. Rev. Lett.*, **5**, 1960. Beautiful. With all the
power and, no less, majesty, of quantum mechanics, it is disconcerting to realize that, as Feynman has said (he died the day of this

writing) we really don't understand the why of it. Some day, perhaps.

In liquid helium, the refrigerant itself, we have the flux quantum analogue in rotons—quantized vortices. "Evidence for the Quantization of Circulation in Liquid Helium II" was reported by Chase in REVIEW *Letters*, Volume 4, 1960. Indeed, the analogue of the AC Josephson effect has been contrived in superfluid helium by Richards and *Anderson* [*Phys. Rev. Lett.*, **14**, (1965)]. In 1947, Tiza (REVIEW, Volume 72) had "The Theory of Liquid Helium," based on his 1938 suggestion that liquid helium below the lambda point could be considered as a mix of two fluids—super and normal. Two years later in Volume 75 of the REVIEW, *Landau* came in a long Letter and, while praising the early suggestion, largely dissented with the conclusions Tiza drew from his theory; Tiza was in rebuttal in an accompanying Letter. Critical work on the atomic theory of liquid helium was that in three papers, I, II, and III, of *Feynman* [*Phys. Rev.*, **91**, (1953) and **94**, (1954)], who considered the liquid at close to $0°K$. In the $E(k)$ *versus* k spectrum, a linear portion rising from $0°K$ shows phonon-like (longitudinal) excitations and, at a minimum in the spectrum at higher temperature, an excitation is found, which is rather like Landau's rotons.

It is not only liquid helium four that has been of interest. With nuclear reactors around after the war, He^3 became available—in small quantitites, but adequate enough for macroscopic experimentation on the liquid. In 1966 a communication in *Phys. Rev. Lett.* **17** from W. R. Abel, A. C. Anderson and John Wheatley reported the "Propagation of Zero Sound in Liquid He^3 at Low Temperatures"; one has come to the point where $4°K$ is pretty hot. As if to emphasize just that point, in 1972 Osheroff, Lee, and Richardson at Cornell presented to REVIEW *Letters* "Evidence for a New Phase of Solid He^3" in Volume 28, following it in the next volume (same year) with "New Magnetic Phenomena in Liquid He^3 below 3mK," two new magnetic superfluid phases, called A and B, in the liquid, having been discovered in the melting curve of the solid co-existing with the liquid. The low temperature physicists are setting up to go down to the μK regime.

An area of concern in condensed matter physics is that of critical phenomena. *K. Wilson* in a notable two-part paper from Cornell in Volume 4 of PHYSICAL REVIEW B (1971), titled "Renormalization Group and Critical Phenomena," applied Feynman's techniques in another unexpected place. Later, *Wilson* joined with Michael Fisher

[*Phys. Rev. Lett.*, **28**, (1972)] in "Critical Exponents in 3.99 Dimensions," a breakthrough of sorts, even unto the understanding of situations in the usual three dimensions.

With 200 feet of PHYSICAL REVIEWs, supplemented by 30 feet of PHYSICAL REVIEW *Letters,* on library shelves, it is obvious that the few of the significant contributions to physics we have cited in the foregoing are but a minute part of what has been published in the journal's second half century. That much of what has been noted is indeed worthy of note, has been the recognition of the Nobel Committee in the award of their prize to many of the authors. There is reason to expect that many more awards will be forthcoming for work both done and already published in the REVIEW and for surprises yet in the offing, one day to be published by the REVIEW or its *Letters.*

While it may seem to have been implied otherwise in the preceding pages, it should be obvious that not all the great discoveries or developments in physics have had first or primary report in the REVIEW or its adjunct, in either the pre- or post-World War II eras. The venerable and widely admired British journal *Nature* has long been used for first announcements in its Letters telling of significant discovery, and it still continues to do so. That journal, going back to its first issue on Nov. 4, 1869 (with Lockyer's famous "Recent Total Eclipse of the Sun"and T. H. Huxley on "Triassic Dinosauria") and not confining itself merely to physics nor only to reportage of scientific work and results, is fairly fast, carries opinion, review articles, news notes, and editorials, serving quite a different function than our PHYSICAL REVIEW. Hansch's (*et al.*) work on the hydrogen fine structure came in *Nature*; Maiman's first laser announcement before that; the structure of DNA of *Watson* and *Crick*; the pulsars of *Hewish* and Bell; Ewen and *Purcell's* detection of the galactic 1420 mc hydrogen radiation; the Brown-Twiss intensity interferometer; many others. *Mössbauer's* narrow gamma ray line came in the *Zeitschrift für Physik*, as did *von Klitzing's* report on the quantum Hall effect and *Bednorz* and *Muller's* ceramic superconductor. *Gell-Mann's* quark proposal came in the Dutch *Physics Letters*; he was of the opinion that the PHYSICAL REVIEW editors would throw it out! *Rubbia* and crew announced the ± W and the Z° particles also in *Physics Letters,* understandable enough that it go to a European journal (and of fast response) with practically the whole of CERN given over to the enormous project. *Penzias* and *Wilson* told of the evidence for the 3°K black body radiation of space in the *Astrophysical Journal*. *Kastler* published his double resonance technique in *Compte Rendus*; and in *Le Journal de Physique et le Radium*, he opened the door on optical pumping. Bell published his inequality in the American short lived *Physics*, and Feigenbaum introduced chaos to physics in the *Journal of Statistical Physics* So the PHYSICAL REVIEW and the PHYSICAL REVIEW *Letters* were (and are) not the only journals one should watch by a long shot, but they have surely had their share of the great discoveries since World War II. *Physics Today* serves a very useful function in spotting for us the important developments, wherever published, especially in its annual January review of progress in various fields of physics during the previous year; a noteworthy contribution to our literature.

But enough. This long chronicle was intended to cover only the first half century of, and the Cornell connection to, the PHYSICAL REVIEW. And there it is, in the first 180 or so of the pages behind us. The reason for this last cursory look at a bit of the second half century of the REVIEW has been to indicate something of the reputation and sophistication the journal had acquired by the end of the first half, if any further proof of that needed demonstrating. The record we have laid out in our selection of papers and authors published during the first period has provided, it is hoped, a balanced view and a feeling for the growth of the REVIEW from birth, through immature youth, to full maturity, which one trusts is not yet that of old age. It is certainly not in age yet of decrepitude. At the same time, what we have seen additionally in the perusal is a glimpse of the growth of physics generally and of American physics particularly, which for the last sixty years has been representative of, and gone hand in glove together with, the growth of physics around the world.

In comparing the second half century of our journal with the first half, one is impressed with the much greater depth of the physics of the more recent period over that earlier, but, more, the enormous quantity being done. One is simply bowled over by the mammoth enterprise the publication must represent and by the amount of library space taken up by the volumes since 1945. The number of entries in the cumulative indices of recent years is flabbergasting. The listing of authors in many single papers is also mind boggling, upwards of ten or twenty frequently, and approaching a hundred on occasion, and from several institutions—quite typical of high energy experiment carried out on large machines. (We are being conservative; in the summer of 1989, in two adjoining communications to

PHYSICAL REVIEW *Letters*, both on the mass measurement of the Z°, one from SLAC and the other from Fermilab, a total of 400 authors was listed.) The record in ratio of authors and their institutions to article content is perhaps held however by a short REVIEW letter in Volume 88, 1952, on the "Experimental Verification of the Relationship Between Diffusion Constant and Mobility of Electrons and Holes." The authors listing was longer than the letter itself, 64 authors from 33 institutions; so long the list that it was given in small print as a footnote and lumped under single authorship as Transistor Teachers Summer School, which was a training course Bell Laboratories organized to get the new technology into university laboratory courses. Single observations involved much uncertainty so a large number of independent observers was important and conveniently at hand in the verification of $D/\mu = kT/q$, the "first direct experimental proof of the validity of this equation for holes and electrons of which we are aware."

The situation is not unique to the PHYSICAL REVIEW or PHYSICAL REVIEW *Letters*; the yearly index of *Science Abstracts* literally gives one a sinking feeling. It seems an impossible situation, great gangs of people somehow have to be employed in culling and assembling it all, not to mention the researchers and authors culled.

There is little point to speculating how publication of the various journals associated with the American Institute of Physics (and others, not only in this country) is to continue. Recently the Institute has added yet another two to its own stable—*Computers in Physics* and *Chaos*. So far as our five PHYSICAL REVIEWs (A, B, C, D, and E) and *Phys. Rev. Lett.*, are concerned, it seems inevitable that further sub-division will come. Will PHYSICAL REVIEW, B, Condensed Matter, be broken down in to B_1, B_2, etc., Metals, Semi-conductors, Liquids, etc., and PHYSICAL REVIEW D, Particles and Fields, into D_1, D_2, etc., Quarks, Strings, and what not? Will the *Letters* be sub-divided? Or will we be receiving printed matter at all? Might not the sub-sections to which we subscribe come to us by electronic means, direct from the author's terminal, to be displayed for us in our offices on a computer screen, recorded upon reception on floppy or compact discs or some such? Where, in that case, the referee? (Where, to put it bluntly, are library systems headed? The American Physical Society along with all other publishers wonders about the inevitable impact of electronic publishing. Now, early in 1994, the stage seems set for the expected onslought of on-line, electronic journals. With all the high technology, it is still somewhat anachronistic that the final assembly of an issue harks back to an earlier time; the typed copy is cut up and pasted—"cut and paste"—on appropriate sheets to be photographed for transfer to lead cylinders used in the final printing.)

On the other hand, can the support of our science grow much further, or for that matter, be maintained for long at the present level? Somehow the endeavor has to be paid for; this has become largely a matter for national governments. A large number of REVIEW papers for a long time have ended with the acknowledgment of appreciation for help from this or that government contract and agency. Will not society see other concerns as of higher priority? One already (1988) sees signs of that in plans for the super-conducting super-collider, high energy machine; in planetary explorations; space stations; mapping the human genome; not everyone is in favor. Will government finance these high flown exercises? (Apparently not all of them; late in 1993 Congress killed the collider, well into its construction.) Will support of such projects diminish support in other less grandiose but perhaps more "important" fields, as it seems must be the case? And if the big projects are not funded, will present support in other fields be continued at current levels? Answers to these questions, and others similar, will largely determine the trend in the growth, up or down, of the PHYSICAL REVIEW. In a way, it would be sort of nice to go back to a journal of size and compass we had at the end of the first half century. An individual could get through an issue and comprehend some of it back in those days, but that is not likely to come about. One must admit, however, that our view today of the physical world is a lot more interesting than it was back then. Someday maybe we will understand it all, and physics journals can be put to rest. At any rate, to what publishing will evolve is probably beyond the imagination, and is anyone's guess. However it goes, the developments will surely impact the PHYSICAL REVIEW and its family in the not too distant future.

And there we leave it.

Printed in the United States
By Bookmasters